D1707544

Grass

ORGANISMS AND ENVIRONMENTS

Harry W. Greene, Consulting Editor

1. *The View from Bald Hill: Thirty Years in an Arizona Grassland,* by Carl E. Bock and Jane H. Bock
2. *Tupai: A Field Study of Bornean Treeshrews,* by Louise H. Emmons
3. *Singing the Turtles to Sea: The Comcáac (Seri) Art and Science of Reptiles,* by Gary Paul Nabhan
4. *Amphibians and Reptiles of Baja California, Including Its Pacific Islands and the Islands in the Sea of Cortés,* by L. Lee Grismer
5. *Lizards: Windows to the Evolution of Diversity,* by Eric R. Pianka and Laurie J. Vitt
6. *American Bison: A Natural History,* by Dale F. Lott
7. *A Bat Man in the Tropics: Chasing El Duende,* by Theodore H. Fleming
8. *Twilight of the Mammoths: Ice Age Extinctions and the Rewilding of America,* by Paul S. Martin
9. *Biology of Gila Monsters and Beaded Lizards,* by Daniel D. Beck
10. *Lizards in an Evolutionary Tree,* by Jonathan B. Losos
11. *Grass: In Search of Human Habitat,* by Joe C. Truett

Grass

IN SEARCH OF HUMAN HABITAT

JOE C. TRUETT

Foreword by Harry W. Greene

UNIVERSITY OF CALIFORNIA PRESS
Berkeley Los Angeles London

The publisher gratefully acknowledges the generous support of the Gordon and Betty Moore Fund in Environmental Studies of the University of California Press Foundation.

University of California Press, one of the most distinguished university presses in the United States, enriches lives around the world by advancing scholarship in the humanities, social sciences, and natural sciences. Its activities are supported by the UC Press Foundation and by philanthropic contributions from individuals and institutions. For more information, visit www.ucpress.edu.

University of California Press
Berkeley and Los Angeles, California

University of California Press, Ltd.
London, England

Library of Congress Cataloging-in-Publication Data

Truett, Joe C. (Joe Clyde)
Grass : in search of human habitat / Joe C. Truett ; foreword by Harry W. Greene.
p. cm.—(Organisms and environments ; 11)
Includes bibliographical references and index.
ISBN 978-0-520-25839-6 (cloth : alk. paper)
1. Grassland ecology. 2. Grasslands. 3. Grasses. I. Title.
QH541.5.P7T78 2010
578.740978—dc22 2009018655

Manufactured in the United States of America

19 18 17 16 15 14 13 12 11 10
10 9 8 7 6 5 4 3 2 1

This book is printed on Natures Book, which contains 30% post-consumer waste and meets the minimum requirements of ANSI/NISO Z 39.48–1992 (R 1997) (*Permanence of Paper*).

If you want me again
look for me under your boot-soles.

Walt Whitman, *Leaves of Grass*

Contents

Foreword

Grass: In Search of Human Habitat is the eleventh volume in the University of California Press's series Organisms and Environments, whose unifying themes are the diversity of plants and animals, the ways they interact with each other and their surroundings, and the implications of those relationships for science and society. We seek books that promote unusual, even unexpected connections among seemingly disparate topics, and that are distinguished by the talents and perspectives of their authors. Previous volumes have spanned topics as diverse as ethnobiology and lizard evolution, but none has so directly and synthetically addressed our relationships with nature as this one.

Joe Truett's poignant, measured prose chronicles the life and times of a class of habitats that once occupied large swaths of North America. He conjures up wall-to-wall blue skies and endless grass, describes the transformative effects of prairie fire and millions of large herbivores. He explores cultural and economic factors that propel wholesale conversion of grasslands for production of cereals and livestock. *Grass* is thus about both our past and our future. At one level it provides scholarly yet easily accessible answers to a series of related questions: What are grasses? Where are they found and why? How do other inhabitants interact with the dominant vegetation? And over the long haul—from bipedal apes on Pliocene savannas, through prehistoric origins of agriculture, to the water-sucking golf courses and manicured lawns of today's big cities—how have we occupied, utilized, and re-created grasslands?

On another level *Grass* unfolds as a heartfelt memoir, a deeply personal yet scholarly meditation. We get a feel for wild places and feral lifestyles through the eyes of a shy country boy roaming the East Texas "piney woods" with a .22 rifle to put squirrels on the family table. We experience the pervasive influence of family and teachers, empathize with his visceral distaste for lawn mowers. Like many a youth, Joe imagines a more exciting outside world, goes to college, and finds cherished beliefs challenged. Later, as he gains the seasoned skills of a professional ecologist, we come to appreciate profound distinctions between tall- and shortgrass prairie, see how grasslands dominate under particular climatic regimes. Along the way he writes with firsthand authority about cattle, bison, prairie dogs, rural values and myths, and the false efficiencies of certain grazing practices. He describes with fairness and bemused curiosity what some regard as the ultimate in conservation initiatives, known as "Pleistocene re-wilding."

At stake here is the heart of one of our preeminent landscapes—what we've lost, what's left, and what we might still have again. Grasslands boast some of the earth's richest soils and, not surprisingly, some of its most productive and thus most endangered habitats. By the end of the nineteenth century, invading Europeans had driven the Great Plains' once prevalent bison to the brink of extinction and severely modified most of that region to support agriculture. Now Brazilians are rapidly clearing tropical savannas for soy and sugarcane, and with them the homeland of maned wolves, giant armadillos, and other endemic wildlife. Joe is unapologetically uncomfortable with our rapacious, shortsighted ways. He longs for a continuous prairie from southern Canada to northern Mexico, for human lifestyles more in tune with what's biologically reasonable. And he makes a compelling case they'd be more sustainable as well as more emotionally rewarding.

Author of an award-winning earlier book on his native East Texas and a respected scientist, Joe Truett works as a restoration ecologist and endangered species manager for one of the United States' largest private landowners. His job is to put the likes of Bolson tortoises, black-footed ferrets, and Mexican wolves back on certain New Mexico ranches. He's got grown children too, worries about their legacy, and asks, how will it all play out? In this wonderfully readable, quietly moving volume, he preaches hopeful skepticism and open-mindedness; he inspires a willingness to learn from nature and human history. In so doing, *Grass* sets the stage for a new conversation about what kind of world we'll live in and with whom we'll share the planet.

Harry W. Greene

Acknowledgments

In writing this book I have built upon lessons learned from a myriad of people and animals. They are far too many to acknowledge here, or even to remember. So rather than trying to recall all who contributed and in which order of importance, I'll mention a few who persist in memory as being especially helpful.

Family comes first. My wife, Judy, consistently steered me toward clarity and away from verbosity. Sandy—Judy's friend and my first wife until she passed away nearly forty years ago—always preferred story over sermon, and I hope she would be pleased. My two sons, Sam and Jed; their respective partners, Carmen and Vanessa; and Jed and Vanessa's children, Geronimo and Jericho, motivated me to carry this book to completion. My mother, Versie, and especially my father, "Boose," who died several years ago, still warn in my memory against believing all you hear and straying too far from the demonstrated truth. My brother, Jack, and his family remain rooted to the homestead where I lived as a child; by example they teach the lessons of habitat and survival.

Other extended family and friends contributed much. Mary Schlentz, Sandy's mother and my sons' grandmother, remains one of my best critics. Mike Rose critiqued the entire original manuscript, to my great benefit. The Coates family—Jim, Nancy, and Milagre—read parts of manuscript drafts and offered constructive criticism. Casey Landrum prepared the line drawings.

My work with the Turner Endangered Species Fund, focusing as it did on grassland animals and their habitats, provided an unrivaled platform for observation and experience. The Fund's executive director, Mike Phillips, an astute ecologist among his other considerable talents, offered unwavering support for and approval of this book. My enthusiasm for the Fund's mission, which arose from the conservation vision of founder Ted Turner, gave impetus for some of the essays that follow.

The University of California Press editorial staff and the manuscript reviewers they selected did an unexcelled job of encouraging, criticizing, and advising. Harry Greene planted the seed for this book and nurtured its development. Blake Edgar skillfully brought his and reviewers' ideas together to suggest changes that greatly improved the original manuscript. Matthew Winfield and Lisa Tauber nudged me along gently. Anne Canright skillfully and tactfully edited the final version. Reviewers James Shaw and Ernest Callenbach encouraged me with their suggestions and compliments, and an unidentified reviewer challenged me with constructive criticisms.

Prologue

After many years of studying wild animals, I am convinced we can learn a lot from them in our quest for sustainable prosperity. As is the case with the other species, habitat limits our abundance and well-being; however, most of us seem to have lost touch with this reality. Fooled by dizzying advances in energy capture and technology that have obscured the constraints of habitat, we have lapsed into a kind of collective hypnosis that denies limits to our numbers and appetites, that assigns us godlike ability to stretch indefinitely the ecological bonds that confined our ancestors.

This culture of denial is pulling us into dangerous waters. Like Icarus of the waxen wings, we go ever higher, seeking the sun, and I worry about the repercussions. This book arose in part from that concern. Some of the convictions that underlie the essays that follow are:

- The twentieth century witnessed the greatest level of average affluence the world's human population may ever know.
- We face a desperate twenty-first century during which the human population will—absent catastrophic disease or war—continue to increase and the energy and other resources that have generated affluence will decline.
- Affluence may drop farther in the United States than elsewhere as others around the world scramble for a more equitable share.
- Declining habitat quality will quicken the fall and diminish the refuges available to the fallen.
- Understanding evolution-based habitat preferences and dependencies can help us and our descendants deal more sensibly with these changes.

Three life circumstances kindled this inquiry into grasslands, a habitat I believe we are genetically programmed to appreciate. First, a rural upbringing gave me early exposure to a world very different from, and in my opinion more satisfying than, the urban and suburban habitats that shape most of today's population. Second, decades of investigating the habitat needs of other animals kept prompting the inevitable question, What about us? Third, I see few educators and policymakers who seem to understand the importance of habitat in shaping our behavior and limiting our numbers.

My two sons and their families count among the anxious contenders for economic security in America. They strive more diligently than I did, and that just to stay in place. Income expectations are high, work competition intense. Many willing to do the same jobs for less pour into the country, while others take over similar jobs that we have exported abroad.

The majority of Americans reside in urban areas. They choose city life because they prefer the cultural amenities, the economic opportunities available, or both. Their children will grow up in a built environment, not because that is the most nurturing habitat but because it is where the American dream has come to reside. It is the place where the past hundred years of cultural experience has guaranteed the best paycheck and the most alluring array of stimuli.

I entertain no evangelical vision that the uneasy occupants of high-rise apartment buildings and office complexes will read this book, pack up their bags, and flock to the rangelands. Most of the recent prospectors for a better habitat seem to have come not from marginal or mainstream economies but from the wealthier classes. People with money can afford to bring city comforts to the countryside and to move away if the going gets hard.

In any event, only a few will experiment with this option—and that is fortunate. The habitats that resemble those in which we evolved can absorb only a small fraction of the masses in today's cities and mega-cities. We have climbed too far out onto the twin limbs of excess—population numbers and per capita consumption—to easily return to the tree of our ancestry. In time, pandemics, wars, and starvation may do for us what we have not had the courage to do ourselves, as we continue to overburden the earth's carrying capacity on the back of subsidies. For now, we are stuck where we are.

But perhaps a few will ask the hard questions. Each year my own small community sees an ebb and flow of the disenchanted. Not all have money; some have elected to stem the cash flow out the back door rather than strive for more coming in the front. The wisest among them listen to the wisest of those who grew up in this habitat. They will be among the survivors.

Like grasses creeping back onto a denuded plot of land, repopulation of the natural human habitat in a way that works for the long term will take time. In Bob Dylan's words, those who were last may come first. The perceptive and the hardy will survive, and tell their children. In this book, I hope only to raise a few questions, plant a few seeds. Those of a later time will testify to the outcome.

ONE Promethean Legacy

Figure 1. First atomic bomb, .016 second after detonation, 1945. Photo by B. Brixner, courtesy Los Alamos National Laboratory.

Energy fuels life. Like other animals we seek the best ways to capture it and funnel it to our own purposes. Our bodies glean energy from the food we eat. Some half a million years ago, anthropologists say, our ancestors gained control of fire to cook their food and warm their bodies. Maybe the ancient Greeks would have pointed to those long-ago campfires on the African savanna as evidence of the original visit by the mythical Titan Prometheus, who stole fire from the gods and bequeathed it to humans.

But that was only the beginning. The next installment came much later—just 10,000 years or so ago. Somewhere around this time, agriculture emerged. By domesticating plants and animals, we began to funnel more of the sun's energy to people. Our numbers mushroomed.

Ten millennia after that, during the nineteenth and twentieth centuries, the Industrial Revolution opened up the way for us to capture even greater amounts of energy. We began to mine and burn fossil fuels, long sequestered in the earth's crust. We prospered and multiplied as never before. Perhaps Prometheus smiled.

Then, on a midsummer morning in 1945, the most concentrated display of energy ever liberated by humans emerged in a remote New Mexico desert. On 25 July at exactly forty-five seconds past 5:29 A.M. Mountain War Time, a light ten times brighter than the sun materialized from the fading darkness. It bathed nearby mountains in a brilliance never before seen, blinding animals that watched from close at hand and startling others in the distance. Ranchers and farmers on their predawn rounds up to several hundred miles away stopped to marvel at the great light looming on the horizon.

At the source, the earth shook. Within seconds, an intense heat fused thousands of tons of soil into glass. Scientists later would say that all life within a mile perished. Shock waves rolled out in concentric circles. Ten minutes later, the ever-widening pulse of pressure broke windows in Silver City, 150 miles to the southwest. Somewhat prophetically, this awe-inspiring release of energy emanated from the southern New Mexico plain called the Jornada del Muerto, or Journey of Death.

The world's first atomic bomb, our ultimate claim on the gift of Prometheus, had exploded. In the months prior, parts for the five-foot-diameter explosive device, christened the "Gadget," had been constructed at the top-secret Los Alamos laboratories. Then Gadget and a 240-ton container called Jumbo had traveled south by train to a lonely railroad siding on the sprawling Armendaris Ranch. Machines unloaded Jumbo onto a massive 64-wheeled vehicle, which crawled eastward across the sage and grass to Ground Zero, known ever since as the Trinity Site. Today you can visit the Trinity Site a couple of days each year when locked gates fall open to tourist access.

Less than a month after Gadget ignited, similar devices named Little Boy and Fat Man brought unbelievable havoc to Hiroshima and Nagasaki. World War II ended quickly after that, and scientists started planning peaceful uses for atomic energy.

• • • • •

I came into the world four years before the explosion of Gadget and the end of World War II. Once the war ended, newly tapped supplies of raw materials, fossil fuels, and nuclear energy, combined with a capitalist arrangement that rewarded inventiveness, propelled Americans into an unprecedented half-century of wealth. My state, Texas, wallowed in oil, another Promethean gift to God's chosen. A unique source of energy important to prosperity then and now, oil catapulted America into the economic leadership of the world. Opportunity trickled down to nearly all.

I grew up in the backwoods of East Texas. Tall trees crowded against our tiny farmstead. Family traditions provided a kind of economic training camp for taking advantage of the wealth conferred by the abundant energy. Parents shaped by the hard times of the prewar depression demanded obedience and considerable effort. Work under prudent guidance could, in that energy-rich time, generate unexpected rewards.

My father introduced me to the theory of relative wealth. If you're well fed and warm, he said, having more things does not make you richer, only snootier. Although he made below the average national income, he felt privileged, I think. Sometimes he spoke of his own father who, still reeling from the Great Depression, had died shortly after my birth. My father sometimes would say, "I wish Daddy could have lived to see these good times."

A peculiar quirk of family dynamics set me off on my career. My father had a tendency to put my brother and me to work if we lay around the house. Not particularly fond of chores, we plotted escape. We called it "going to the woods." Now I see it as a temporary retreat to the Stone Age. Or more appropriately the Stick Age, there being few stones in our habitat, which many called the Big Thicket.

As a cabinetmaker, Daddy worked away from home during the day, returning in the evenings. Outside our house, within a thirty-second sprint, great beeches and oaks could screen us from view, and when it came time for him to come home we often melted away into the woods. Eventually he built his own shop beside our house. After that, during weekends and school vacations, we sometimes kept to the woods all day to avoid having to sand cabinet doors, add to the store of winter firewood, or weed the front pasture.

Two decades earlier, in leaner times, my father had been forced to drop out of school in fifth grade to help support the family by farming. He never went back to school. We had it easier, my brother and I. Only much later did I realize how much easier, and why.

We played Tarzan and Jungle Jim, discovering a life that our genes recognized as proper. Mama subsidized the vagrancy. She packed sandwiches for longer jaunts and patched shirts we'd torn from overestimating our abilities to leap from branch to branch. Granddaddy and Grandma, who lived up the road the length of a football field away, showed us how to navigate the woods. They took us to dewberries in season, secret plum trees near abandoned homesteads, and fox grapes hanging low on the vine in the pinewoods hills. We had space, my brother and I, a privilege now lost to most Americans and indeed most humans. Woods stretched endlessly, the biggest chunks open to our explorations because of absentee owners, or tolerant ones.

For one dollar a year Granddaddy leased from Kirby Lumber Company three thousand acres of open pine forest a few miles beyond the hardwood habitat that closed in around our house. On this tract he ran cattle and wild woods-hogs. Once we grew old enough and gained possession of twenty-two–caliber rifles, we followed him. We gloried with him in his cows, which he called to feed with a far-reaching yodel, and we helped him harvest two-hundred-pound hogs that he'd rounded up as piglets and earmarked—the standard way of claiming ownership of free-ranging pork.

I decided to live this way forever. Eventually, and against all apparent odds given the history of industrialized man, it began to seem possible. As high school neared its end, I discovered that Texas A&M College, three hours' drive to the west, offered a curriculum in wildlife management. My parents and teachers did not seem to care what the subject of study was, only that I break the local code and go beyond high school. I built a dream of the future: to get paid to return to the woods and live among the animals.

• • • • •

Taught for eighteen years the credo of rural East Texas, I had no notion when I left home that my values would change. Gradually I circled farther and farther from the security of home, entertaining new notions, questioning the old. Changes came slowly for me, imprinted as I had been onto a primordial existence in the woods. And unlike some on the journey from home, I clung to the most basic beliefs that instructed survival. One in particular that my father taught has stayed with me through the years: *Be careful who you believe.*

He is dead now. In retrospect, I see him as a natural skeptic, never fully comfortable with organized religion, the Freemasons who drafted him later in life, or the herd behavior of neighbors at the polls. *Be careful who you believe.*

One of my teachers in high school and the best ones at university phrased my father's philosophy another way: Question dogma. To question dogma in my younger years would have contradicted the obedience that worked well then. Probably obedience was a good arrangement for us youngsters. Given the risks that hovered over kids running wild in the woods, doubting the adults wasn't the best way to survive. But as I get older, dogma seems more and more in need of questioning.

• • • • •

In wild animals, there comes an age when the young get restless. Their blood chemistry changes, their hormones surge, the closeness of birthplace begins to suffocate. The offspring, particularly the males, fall out with the parents. They leave home, sometimes at high speed with the elders in chase. Biologists call this dispersal. By no accident, it coincides in time with entry into sexual maturity.

The human animal proves no different. We become restless as our bodies blossom in the teenage years. Daughters squabble with mothers; sons squirm under the rule of the old man. We leave if we can afford it, searching for new meaning and like-minded strangers, particularly those of similar age and opposite sex. The dispersal urge hit me especially hard.

In the wide world beyond home lurk those who would prey on dispersing youth. Army sergeants love teenage recruits—malleable, impressionable, looking for a place to fit in. The magnetism of universities and the social unrest they generate likewise reflect the search for the new, the rejection of the old. The search does not always proceed sensibly, said the late philosopher Eric Hoffer, who after World War II saw many young people become True Believers under the tutelage of social and religious extremists. Being born in a fortunate place and circumstance, I avoided the less desirable of the evangelists and at dispersal age set off for college.

TWO Out of the Forest

Figure 2. Western movie poster, 1928. Courtesy Yale Collection of Western Americana, Beinecke Rare Book and Manuscript Library.

I had little admiration for grass when the professors at Texas A&M started serving it up in class. By the time I left home I'd had enough of chopping Bermuda grass from rows of corn and beans, trimming carpet grass and crabgrass with a 1950s reel-type mower, and swinging a weed cutter at the really tall stuff. Most of all I hated the push mower. Exasperation is the word that best describes following this beast through rank grass at the bidding of country housewives who paid only occasional homage to the social code of Short Grass in the Front Yard.

The woods always beckoned from beyond garden and lawn. They offered squirrels to hunt, mysteries to explore. Dull blade of hoe and mower coupled with the maddening vigor of grass watered by fifty inches of rain per year did little more for me than lay out the lessons of patience.

Grass did help fortify farm life, in a vague sort of way. Our Jersey cow ate it, and I loved her cream and milk. The pair of sorrel geldings my grandfather used for plowing in the spring and rounding up woods hogs in the fall lived mostly on grass. But at that time, ecological connections more than one step removed from pleasure escaped my interest. Grass represented work and that's all I needed to know.

Early unease with grass matured into full-fledged distaste that first year at college. Texas A&M required two years of service in the Cadet Corps, a college version of military boot camp. With my classmates I endured morning inspections, three-times-daily marches to chow, and Saturday drills that pretty much shot the chances for hitchhiking home for the weekend. Fresh-mowed Bermuda grass tainted the air along the paths of march—"Hut, two, three, four! Here's the way to win this war!"

Late in the first year I found that those of us enrolled in wildlife biology would be required to study grass. My faculty advisor said, "Look here, we'll sign you up for plant ecology and range management. You'll study agrostology—that's the science of grass. You need to know what sustains large animals, where red meat comes from."

I groaned internally, thumbing in vain through the course listings in search of Animal Trapping 201. But I did not object, for, back then, professors sat at the right hand of God.

Grass gradually elevated itself in my esteem, imperceptibly at first. My history professor called it the lifeblood of the great cattle drives from Texas north following the Civil War. I began to notice it as the occasional backdrop that glorified Randolph Scott and John Wayne at the western movie I afforded once a week. I saw no thankless work on that screen, but only endless horizons and pretty girls chasing young men across the prairie.

In 1962, with the two mandatory years of military training over, you could find me on Saturday afternoons behind my civilian dormitory. With West Texas classmates I practiced roping a sawhorse adorned with a fake cow's head cut from a Dunlap tire. Whirr, whirr, whack! Whirr, whirr, whack! Snap of the wrist to tighten the noose. Retrieval of the manila coil with cool inattention.

Occasionally a student who owned a car would drive us to the western clothing discount store in the village of Lott, up toward Waco. There we found trim-cut shirts with shiny snaps in place of buttons, and Tony Lama boots far beyond my means. I bought a cheap western hat. Back in the dormitory, the cowboys down the hall hooted, advising me to try a smaller brim.

And so, at this university on the eastern edge of the Great Plains that stretched from Mexico to Canada, I began a lifelong delving into the great mystery. What was it about the prairie, the great expanses of grass, that people found so alluring? That freed them from the bondage of toil, not by compelling less work, but more? That elevated the herdsman, the cowboy, and the ranchwoman above the shadow of the man with the hoe and the woman with the broom?

• • • • •

In my junior year I took plant ecology from a professor in love with grass. Early in the class he told us how grass produced most of the calories consumed by people. "Look at the basic foods of nations," he said. "Grass seeds—rice, corn, wheat, millet, barley, oats, and so on—make up the great majority. You may not think that noteworthy except that, of the world's 400,000 or so species of vascular plants—the ones with circulatory systems, like trees, bushes, wildflowers, weeds, and grasses—grasses account for only about 8,000, some 2 percent."

He paced as he talked. "We refine sugarcane, another grass, to make pecan pie sweet, people diabetic, and the Imperial Sugar Company down by Houston rich. If you like beef, ride horses, or rope steers on weekends, you need to know they all came from grass. And remember that it's grasses, not grass, when you're talking about more than one species."

A few weeks after the course began, the class loaded into a college bus and drove west. "We're going to see a remnant of tallgrass prairie," he said. I still remember how his voice lifted when he announced it, as if freshened by the thought of getting a break from classroom air, unconditioned back then and heavy with late-September humidity and farmboy odor.

Ten miles down the highway we came into the Brazos River Valley. "Germans and Czechs settled this area 130 years ago," he shouted over the noise of the bus engine. "Now it's part of the Texas A&M system." Looking out, I saw fields beside the road displaying signs of university ownership: bags enclosing corn tassels, experimental replicates with crop varieties named on placards, students crouching over study plots. We traveled on, up and out of the floodplain and west.

In cryptic bursts the professor described how this part of Texas used to be, before the coming of the cowboy and the exodus of the Indian. A sea of grass, clear of brush to the horizon, dotted with buffalo. When he swept his hand toward the window, I looked out past the spotted cattle in the roadside pasture and glimpsed them there, just for a moment, those wild people chasing the wild animals of the olden days.

Then the bus lurched off the pavement and stopped beside a field of grass. The plants, unusually tall, swayed in the breeze, looking out of place among the pieces of plowed ground and the weedy pastures, sprung up from previously plowed ground, that surrounded it. "This pasture has never seen a plow," said the professor, "nor been grazed much."

He led us out into the meadow, armed with measuring tapes, surveyor stakes, and clipboards. "Now we're going to measure grass," he said. "We're going to measure canopy cover and basal cover, by the line-intercept method." We fellows looked blankly at each other (there being no girls to look at in 1963 at Texas A&M). But we listened as he pointed out the different species, for we knew they would show up on exams.

"Little bluestem, this grass that comes up to your waist, that's the dominant here. Dominant—meaning there's a lot of it, with a strong influence on the other species." He pointed as he lectured. "And over here Indiangrass, *Sorghastrum nutans.* You'll want to remember that name." He spelled it. We scribbled on our pads. "And oh, look at this big bluestem." He caressed a clump of shoulder-high shoots with tips that looked like chicken feet. "This used to dominate the lower places in the early days," he said, "before your ancestors and mine brought cattle, plows, and a dirt-grubbing German's way of living."

Soon he had some of us stretching the measuring tapes between surveyor stakes. Others leaned over, calling out the lengths of tape intercepted by grass crowns, and yet others jotted the numbers on forms distributed earlier. I joined those crawling along the ground and trying to figure out where a grass "base" started and where it ended.

Once finished with the measuring, I wandered about the meadow, letting seedheads of the grasses slide through my fingers. Almost certainly I had seen some of them before, in the pinewoods pastures of my earlier years. But I could not be sure, for my attention had been on the raccoon tracks and the pine cones dropped by squirrels. Now, though, the grasses had names, and I would remember them.

Students seldom take field trips now, I'm told. I wonder how the fates of nations will change when at last we bow to the governance of those who never went out with a teacher in love with grass. Who never came back with dirt on their knees, uplifted despite themselves by the process of seeing, touching, and smelling plants from unaccustomed postures.

THREE Science and Faith

Figure 3. Science and faith, uneasy companions. Photo by author.

At university I ran head on into the contradiction between science and faith, something I cannot recall being an issue before that. I attended church very irregularly in grade school and almost never in high school, I think now because of my father's mistrust of rural Baptist preachers. Only one teacher at my high school had dared venture onto the treacherous ground of evolution. That state of blissful ignorance ended at Texas A&M.

One memorable incident took place the first day of a beginning class in ecology. After the professor had described how we would study plants and animals from an evolutionary perspective, a student raised his hand and announced that his church didn't believe in evolution. I was dumbstruck, accustomed as I was to respectful attention in class even from dissidents. My marveling continued when the professor invited the student either to drop the course or keep his opinions to himself.

Intrigued, I took a course on comparative religions, which complemented the campus church services I'd taken to attending with my Southern Baptist roommate. On Sundays, preachers held forth on the merits of faith. During the week, professors of biology hammered home the lessons of logic.

I liked the feel of logic. It made sense of the strategy I'd used for hunting squirrels and evading chores. But in classroom and dormitory, on the military drill field, and in the mess hall, my compadres seemed to swallow science and religion simultaneously without choking. I decided to compartmentalize the two for the time being, hopeful that a miracle eventually would reconcile their differences.

In advanced courses we learned the tenets of the scientific approach. Identify the source of the information—that was one. The vague "they say" I'd previously taken for authoritative would not do. Another emphasized replication as the cornerstone of science: drawing reliable conclusions requires repeated observations, not just one. There went most of my grandfather's pronouncements about where to find wild hogs, which tended to be based on his last previous hunt.

"Show me the evidence," scribbled the professor on the margin of a paper I threw together late one night without access to the library. "Let's see the references. Where are the data? And remember that the word *data* always needs a plural verb."

I began to see that my father had tried to teach me science on the basis of his fifth-grade education. He had called it common sense.

At the campus church my roommate and I usually attended, the minister and Sunday school teachers offered the Southern Baptist version of Christianity. "Read your Bible," they said. "It has endured for two thousand years and you can take our word that it is the absolute truth. Those who attend other churches, as long as they are Christian, have good hearts but flawed interpretations of what the Bible really says. Non-Christians need saving, and that's part of our job."

Stories in the Old Testament, most of which I'd never read, began to fascinate me. The power of simple prose, of earthy parables, lured me into and through Genesis, Exodus, and other epics of human fallibility. The New Testament, meanwhile, discouraged me from doing a lot of things that could have gotten me into trouble.

The most interesting part of Christian dogma was the promise of life after death. To arrive at this posthumous experience you not only had to believe it existed, but you also had to know how to get there. Grace, Rapture, Heaven—the Ultimate Destination seemed to have a lot of names, and all the preachers said it was up, not down. One made it there through a succession of triumphs over evil. Although backsliding to lower levels often happened, with proper behavior you could regain lost ground and eventually arrive at that city of gold.

The hymns I liked most pictured Heaven somewhere beyond a river. "Shall we gather at the river, the beautiful, beautiful river. . . . " I could see parallels between fishing on the banks of the Angelina near my home and the approach to Heaven.

Life after death seemed like a good notion, and devoting at least part of my time qualifying for Heaven seemed a small price to pay. Although science and logic gave no indication that Heaven existed other than in the mind, knowing that television and black holes would have been thought preposterous only a few generations before kept me open to the possibility of a logical explanation for such illogical beliefs.

• • • • •

Most Texans knew A&M as the premier agricultural college in Texas. Number one in the world, according to its administrators. It attracted specialists in the production of human food. Sometimes these specialists—our soils and crop sciences professors—gave us their agenda up front. Sometimes they seemed to assume we knew it or we wouldn't be there. Briefly it said: The holiest ground is that which feeds the most people.

The central tenet of range management fit that of the other production sciences: good range is that which produces a lot of cows, and sheep and goats if you're talking about the Texas Hill Country. The professors assured us their beliefs arose from scientific observation and experimentation. But some delivered messages about good and evil on the landscape with the same undocumented assurance I'd seen in Baptist ministers interpreting the New Testament.

By the time I finished my courses in plant ecology and range science, I'd been indoctrinated in the gospel of land management according to one Frederic E. Clements. This towering figure of science, together with his equally influential student John Weaver, had authored my textbook *Plant Ecology*. Both had taught at the University of Nebraska, studied the surrounding prairies, and on this basis developed a theory of community ecology that resounded worldwide.

A plant community resembles in some ways a community of people, my professors said. It consists of plants that coexist and interact. Weaver, Clements, and my professors of range ecology taught that the particular mix of species in each type of plant community—shortgrass prairie, tallgrass prairie, long-leaf pine forest, and so on—had evolved more or less together for thousands of years. Because of this, the species in each community belonged together.

More than half a century before I studied plant ecology, Clements had begun to elaborate a theory he called community succession. We learned about it in class. It reminded me of the Baptist preacher's story of human ascension to Heaven.

In the beginning, said Clements, each plant community started from bare ground. I equated bare ground with the bare slate of the human spirit at birth. Over time, he continued, the community advances through a series of "successional stages"—changing but predictable mixes of species—until finally it enters into the "climax." Disturbance can set it back along the way. Clements called such setbacks "retrogression," which I saw as a kind of backsliding. With proper management, however, you could get the community back on course to climax.

If you don't know much about plant ecology or range science, you may not have heard of Frederic Clements. But almost certainly he will have affected your life if you have ever gloried in a sea of Great Plains grass rippling in the wind, visited a western national park, or hiked across public domain anywhere in the United States. Ever since Clements's gospel of succession took root in the textbooks of community ecology, most scientists managing American landscapes have gazed across their respective domains and dreamed of reaching climax.

Many students' eyes glazed over when professors explained succession theory. I had an advantage, though, having witnessed with my own eyes what the textbook called "old-field succession"—the march first of weeds, then of broomsedge, bluestem, and other grasses, and finally of loblolly pines onto abandoned fields in East Texas. It happened on my grandparents' left-behind land after they stopped farming the old home-place and moved nearer to us. Granddaddy had called it "pines taking over."

• • • • •

Many ecologists in Clements's day believed that nature strived constantly toward climax, the ultimate endpoint, guided by an Invisible Hand or Divine Purpose. Philosophers have called this view "determinism"; some people now might call it Creationism. At the height of Clements's influence in the mid-1900s, ecologist Frank Egler criticized the unquestioning acceptance of succession theory: "We find . . . as in revealed religions—a brilliant and elaborate superstructure raised not as hypothesis but as established fact." Where were the data?

The leading science pundits of the time silenced Egler and other critics by ignoring them. Out on the range, Clements's views have prevailed for most of a century, and many land managers still confidently echo his theme of good, bad, or ugly depending on nearness to climax. To me at university, and I think to many others, his theory represented a kind of odyssey at the community level. It was the kind of story I wanted to believe.

Grassland managers kept the faith with succession theory partly because managing by its tenets brought in more money. In general, the nearer the grassland approached Clements's idea of climax, the more pounds cattle usually gained each year and the greater profit the owners realized. Could it have been that an agricultural mindset motivated Clements and colleagues to look for an ecological theory that would justify profit? Perhaps only Clements and Weaver could answer this question, and they are dead.

Inventing an attractive story to justify a hidden agenda sounds more like politics than science. It seems doubtful that Clements consciously built succession theory to justify utilitarian ends. But a skeptic might point to the Invisible Hand of Bad Assumptions. Believing that humans are the apex of Purposeful Creation might well tempt one to develop theories supporting that belief.

Once I visited the Nebraska State Capitol in Lincoln. Murals of homesteader wives in bonnets, great-muscled men breaking prairie sod, and trains coming to haul away produce domed high above the rotunda and foyer floors. Agricultural icons of the Cornhusker State loomed large on the walls. Over all hung the aura of a Manifest Destiny guided by the hand of the Creator.

The University of Nebraska, several blocks away, had spawned Clements and succession theory. It sustained Professor Weaver his entire career. Having spent four formative years at another agricultural institution in the American midsection, I now find it easy to imagine that food-production scientists could be influenced just a little by biblical notions of Man as divinely appointed Custodian.

• • • • •

Some of my professors in wildlife biology at A&M did not talk much about Clements. Preference of deer, quail, turkeys, and other wildlife species for successional—not climax—communities might have had something to do with it. But neither did they reject Clements's view that the untrammeled pristine, the supposed climax for most communities, represented the ultimate in conservation goals.

Meanwhile, the agricultural people at A&M justified their work with one near-holy cause that never made long-term sense to me at all: We shall feed the world.

In ecology class we learned that numbers of animals and people—unless controlled—grow to fit the food supply. Did the starving millions the A&M professors proposed to feed show any sign they would control their numbers, thus curtailing an endless spiral of more food, more people? I saw no evidence of such control around me, though I heard that it was happening in China. But nobody at A&M seemed to think China's heavy-handed approach to population control was a good idea.

Curious about what looked to me like short-sightedness, I read more. I found that Thomas Malthus had worried about the same thing two hundred years ago, and some leaders in China millennia before that. Usually nobody had listened, at least nobody who could do much about it. Things are different now, said some of my A&M teachers. We have the technology for convenient birth control in humans.

It has been several decades since I graduated from A&M; the world's population has doubled in the meantime, and recurrences of famine, food aid to forestall famine, and food riots forewarning of famine seemingly quicken. The agricultural people in my country remain unruffled. They now assure us they can not only feed the growing millions, but also propel them about in cars running on crops and crop products—biofuel from corn and sugarcane.

To some true believers of a fundamentalist mindset, the threat of world hunger didn't seem to matter in my college years, and it still doesn't. The deserving ones will be saved by the Second Coming, or some other form of transport away from the problem. That belief has its attractions. Maybe I got recruited by the wrong evangelists after all.

FOUR Playing God

Figure 4. Mining coal, Black Mesa, Arizona. Photo by author.

In 1972 near Fairfield, Texas, I got a chance to play God and some money to do it with. The drama started innocently enough. The Industrial Generating Company, a subsidiary of Texas Utilities, had started building the Big Brown Steam Electric Station several miles east of Fairfield. A seven-story generating plant neared completion. Designed to be fueled by lignite—a low-energy version of coal to be mined nearby—it soon would generate electricity to feed into the Texas power grid that electrified homes, offices, and industries throughout the state.

The U.S. National Environmental Protection Act, or NEPA, had become law two years earlier. Stimulated in part by NEPA and the growing interest in environmental protection, Big Brown's senior environmental officer, Dick White, had commissioned pre-mining inventories of soils, landscapes, and ecosystems. What the studies found would guide restoration.

I took on the task of plant and animal inventory. Dick put me up in a spanking new environmental research laboratory with residence quarters. The lab nestled beside the newly filled Lake Fairfield, the waters of which cooled the generators. The plant with its two smokestacks hogged the skyline to the north. In the years since the Texas A&M wildlife and range professors had introduced me to ecology, I'd been away to places with shorter grass and longer views. Now I was back at A&M, this time as a member of the faculty.

Others showed up at the Big Brown laboratory. Elsie, a Ph.D. student from the University of Texas, moved in. She began to study the fish in Lake Fairfield, with part-time help from her husband, Alan, a theoretical physicist at the same university. In early summer a new handyman, recently graduated from Teague High School fifteen miles away, showed up. Ed Walton would mow the grounds, maintain the surroundings, and help us when needed. Ed soon showed me what I've seen time and again since: that advanced schooling confers no exclusive claim on common sense.

One evening after a day of driving, counting birds, and measuring plants on land scheduled to be mined, I joined Elsie, Alan, and Ed in our bid to play God. Convened at the laboratory dining room table, we began to discuss reclamation. Being the first at the scene of destruction, we felt a kind of omnipotence. What kind of firmament should we direct our anticipated disciples—the bulldozer operators, the grass seeders, the tree planters—to create?

"Hey," said Ed, "let's play a game. I'll describe three landscape options. You three tell me which one best appeals to you as a place to build a house or live in. That'll be a good way to start our vision."

He commenced. "Number one, envision a prairie. Imagine grass stretching to the horizon on all sides—like one of those coastal bermuda pastures between here and Fairfield, but endless. Two, think about living in a grove of trees, looking out onto an expanse of grass. Three, imagine yourself in a thick woods, view restricted all around to less than a tennis-court length. Like that post–oak-hickory thicket out by the lignite haul road."

We all voted for the house in the grove, looking out on grass. Plus a little water, maybe a pond or creek. Based on that vision, we saw a Brave New World taking shape at Big Brown once machines had hauled the coal away and refilled the pits: an open savanna with scattered bushes, groves of trees, and stock-ponds with great blue herons, red-winged blackbirds, bluegills, and bullfrogs.

But could we really influence what happened once the draglines and bulldozers had finished their job and moved on? As it turned out, we could. The "overburden" soils, mixed and replaced, proved as productive or more so than the original topsoils. Mixing the original hardpan, degraded by a century of careless farming, did more than fertilizer to effect improvement. Dick White intervened as our mainline to God—he had pull with the company managers. Our preferred landscape began to take shape.

On occasional visits later on, I saw the miracle of landscape rejuvenation in stages. Once the coal was removed, the soil overburden replaced, and the land recontoured, revegetation commenced. The grass of choice for initial replanting, an "improved" strain of bermuda called coastal, had to be propagated by "sprigging," a process that involved dispensing stem segments that later took root. Native grasses came later, both by purposeful seeding and by natural seed-drift from unmined areas. Transplanting small shrubs and trees from suppliers proved best for jump-starting the savanna aspect, and soon there rose up woody copses and travel corridors for wildlife. Waterbirds, raccoons, deer, and other early-successional animals began to move through and recolonize.

And then, dismay. Once Texas Utilities returned the land to private hands, as per the original plan, the new owners more often than not brought in tractors with heavy-duty bush-hogs and mowed the young woody plants to the ground. Why? Ed had moved on by then, but he would have had a simple answer: bushes and trees take up space where cow food could grow.

This is cowboy country, Ed had instructed me that first year. People make money, or hope to, by growing cows. But they don't live out on the reclaimed spoil; they live in Fairfield or Dallas. The reclaimed land isn't their habitat; it's their cows' habitat, their source of cash. Their status goes up and down with the number of cows they run, the money they make, the size of rig they drive. They're not too concerned about scenic qualities of the range.

"Let me give you an example of how powerful the cowboy image is around here," he once said to me. "We had a new kid move into Teague a few years ago. Rich kid. Old enough to drive, and pretty soon his daddy bought him a brand new pick-up. Big V-8. He bought a hat and boots, started hanging around the Dairy Queen, trying to fit in. Most of us, being the country bigots we were, ignored him."

Ed waved his hand, warming to the story. "So he went out and bought a goose-neck trailer. I don't think they had any cows at all, him and his daddy, much less enough livestock to warrant a goose-neck. Anyway, pretty soon you could see him driving around town pulling his trailer, dirtied with manure from a borrowed cow to make it look used. And not long after that he had a girl riding with him, one of the best-looking in our school. It's the image."

Ed proved to be a jokester, and I never knew how much to believe his stories. But he did understand why the perfect human habitat we envisioned didn't make it to the Seventh Day at Big Brown.

FIVE Pleasing to the Eye

Figure 5. The Rocky Mountains by Albert Bierstadt. Courtesy Yale Collection of Western Americana, Beinecke Rare Book and Manuscript Library.

Can science measure what the eye finds pleasing? Artists generally say no. But in a way, that's what I set out to do at Big Brown before the bulldozers razed the landscape. Those planning restoration wanted to understand what features of the habitat attracted animals. With this information, reclamation specialists could bring back the animals by restoring the contours, soils, trees, and grasses that had attracted them in the first place. So in a sense I needed to measure what the various species found pleasing to *their* eyes or, for some, their noses or ears.

Into the field I went, carrying the paraphernalia for measurement. I mapped the plant communities and within each calculated the average sizes and densities of trees, the extent and variety of shrubs, the canopy coverage of weeds and grasses. On these plant community maps I systematically laid out locations for transects and stations at which I would take "samples" of the animal populations, which usually involved calculating their densities or indexes to their densities. I identified and counted birds at "listening" stations, censused large mammals by observing their tracks on roadside strips of sand, and live-trapped small and secretive ground-dwellers.

I jotted down sightings of mockingbirds and painted buntings singing from woodland edges and isolated trees. I scribbled notes about white-tailed deer and bobwhite quail at pasture-woodland margins. I watched killdeers come to barren ground and trapped cotton rats in tallgrass swards. Wood ducks and beavers lingered in the dark bottom woods of Pin Oak Creek. After several weeks of observation, I added up the numbers and described which kinds of animals preferred forest, savanna, meadow, marsh, brushland, or edges between two types.

The measurement gave me something to go on, but a still larger question loomed. How does each species know its proper place, its best habitat? How does it know where to establish its territory or build its nest? University training provided an evolutionary answer: That's where it survives the best. But how do they know which places are best for their survival? The artist might gloat: We cannot know. The scientist might counter: Presumably by some hard-wiring in the brain, some neurological signal stimulated in many species by vision, that we cannot yet measure. Both might agree that some features of the habitat are particularly pleasing to the animals in question.

. . . . .

Future historians may look back on the 1970s as the crowning glory of Manifest Destiny. The decade opened on unprecedented good times in the United States and indeed the entire industrialized world. Congress had signed off on the National Environmental Policy Act (NEPA) in the waning days of 1969, a gesture that signaled more than any other the readiness of Americans to end their fixation on quantity and begin the quest for quality. Powerful pieces of legislation to protect water, air, soil, rare plants, and endangered animals came down the following decade. The rest of the developed world began to follow suit.

People in urban environments had for many years been making money studying the kinds of habitats preferred by the human animal. They designed settings for houses, development complexes, golf courses, parks, and cities. They claimed insight into the ideal human habitat. They called themselves landscape architects.

The same year that Congress approved NEPA, an unusual book started circulating among landscape architects. Arising from the same prosperity that emboldened people to ask about human impacts on animals, it urged architects to look beyond the conventional in planning human habitats. The book, bearing the title *Design with Nature,* came from a Scotsman turned American named Ian McHarg. He has since been compared with Frederick Law Olmstead, the undisputed father of American architecture, the designer of the Capitol Building grounds in Washington, D.C., and the planner of Central Park in New York City.

Ian McHarg grew up in the countryside ten miles from the industrial grime of Glasgow. During World War II he served in a parachute brigade of the British Army. From this aerial vantage point, he saw the devastation of human habitats resulting from the pursuit of money and power. After the war, he came to the United States and enrolled at Harvard University, where he received graduate degrees in landscape architecture and city planning. Subsequently he founded the Department of Landscape Architecture at the University of Pennsylvania, with the determination that architects could do better.

Inviting architects and planners out into nature, the natural human habitat, McHarg challenged them to call forth in their designs "the power of sun, moon, and stars, the changing seasons, seedtime and harvest, clouds, rain and rivers, the oceans and the forests, the creatures and the herbs." Humans, he said, should not live like rabbits in cramped warrens but as "co-tenants of the phenomenal universe . . . participating in that timeless yearning that is evolution."

• • • • •

In the spring of 1972, I drove down toward Houston from Texas A&M with Dr. Jim Teer, head of the Wildlife and Fisheries Department. We had an appointment with some people who were building a city from scratch; they wanted some advice about incorporating wildlife habitat into their town. Texas oilman George Mitchell, with the help of a $50 million loan under the Federal Housing Act of 1970, had put up the cash to initiate the development. Construction had begun. A famous architect up north had conceived the idea and provided the design, the developers said. A fellow named Ian McHarg.

When we visited McHarg's "Woodlands," much of the landscape remained a thick hardwood forest similar to some I had explored as a kid but absent the hills. A few office buildings and residences had been completed. The on-site architects showed us how they were configuring buildings to embrace trees and where they thought green corridors for wildlife should go. The polished oak banisters that steered the way to the second-floor meeting room bespoke the anticipated clientele. The development already housed a few thousand residents, they said.

Now, some forty years later, the Woodlands is home to around 60,000 people. Its country club bills itself as the only club in the nation to have six world-class golf courses. I have not been back, but the photographs on the advertisements show the Woodlands to have a lot more grass than it had in 1970.

The moderately wealthy who came to inhabit the Woodlands wanted more grass. They wanted short grass, specially developed for putting greens, fairways, lawns, and streetside embellishment. They wanted trees to be widely spaced to let in sunlight and open up panoramas. They wanted human habitat.

When Ian McHarg died in 2001, obituaries counted the Woodlands among his greatest achievements.

Small clouds loom over the dynasty that built the Woodlands. Oil production in Texas and the United States peaked about the same time Jim Teer and I gave those architects advice on wildlife habitat, and it has steadily declined since. According to the experts, the world's oil production is now near its peak and likewise will soon begin to decline. The remaining store will be harder to get and more costly.

For the average person, Ian McHarg's dream hovers uncertainly. Woodlands-type cities for the masses seem less and less realistic as our economy reels from higher oil prices and alienated producers overseas. Our people and our government borrow with an abandon that would have shocked my Scots-Irish grandfather. How long can we maintain the fiction of promising better and better habitat for humanity, more and more acres of manicured grass? What would Ian think?

• • • • •

Hardly had the ink dried on McHarg's book when other professionals began to broadcast their own versions of the perfect landscape. Anthropologists, sociologists, psychologists, and ecologists all took a shot at it. An idea whose time had come, the search for our lost habitat soon reverberated around the world. Our unprecedented wealth, and the freedom of thought it fostered, gave people new license to explore this very old idea.

One of the dictums of wildlife ecology holds that each species prospers best under a unique set of habitat conditions not shared exactly by any other species. But all share a generally similar need: to eat and not be eaten. The best habitats for each offer less risk and more food than other options.

In terms of basic needs, *Homo sapiens* claims little distinction from other animals. Clearly, we need food. But our historical freedom from predation has obscured what those studying our prehistory know well—once we were prey. *Man the Hunted,* a recent book by anthropologists Donna Hart and Robert Sussman, guesses that early humans probably spent a lot of time figuring out how not to become food of lions and hyenas.

In 1975, British sociologist Jay Appleton published a perceptive analysis of what people look for in landscapes and why. He based his notions on the idea that, deep in our genes, there lies the tendency to select the kinds of habitats that during our evolution would have let us eat more and be eaten less. In his book, *The Experience of Landscape,* he lays out a theory of human habitat preference according to the *prospects* and *refuges* offered.

Appleton recognized first of all that a person's preference for landscape type depended partly on past experiences associated with particular places. A golden childhood—like mine—in a deep forest could instill a love of woods, or a bleak experience in prairie could seed the fear of open landscapes. By conducting experiments to weed out the effects of past individual experience—for example in small children or in groups of people with varied habitat histories—he thought he found signals of the ideal human habitat.

People by and large liked places with good visibility, Appleton found, but with ready access to secure places. In evolutionary terms, the good visibility—from hills and low cliffs in open country—provided the prospect, allowing people to scan for large mammal prey, food-rich habitat, big predators, and human enemies. The secure places—trees, rock outcrops, cliffs, caves—would have provided refuge from big predators, cold rain, and hot sun. Deep within each of us, Appleton concluded, resides a habitat preference originating from the kinds of prospects and refuges available to our distant ancestors.

Close on the heels of Appleton's book, two Smithsonian Institution psychologists, John Falk and John Balling, began testing human habitat preferences. They asked their test subjects to rank a series of landscape images projected onto a screen. People chose, from most to least preferred: savanna, deciduous forest, coniferous forest, tropical forest, and desert. Kids aged eight and eleven—less likely than older people to have developed preferences based on previous experience—preferred savanna most strongly.

In the 1970s evolutionary biologist Gordon Orians of the University of Washington also began researching habitat preferences of people, later teaming with psychologist Judith Heerwagen to test people's responses to paintings of various landscapes. They took particular care to weed out the influence of their subjects' previous experiences. What they found supported the ideas of McHarg, Appleton, and Falk and Balling. People preferred savanna-like landscapes with vistas of grass and scattered trees and shrubs. Prospect and refuge potential explained a lot. People liked scenes that, in a primeval setting, would have promised something to eat and security from being eaten.

Orians and Heerwagen also found gender differences in preferred habitats. Woman, the nurturer, focused on near-at-hand security for herself and her children—short grass, climbable trees, water, and bright colors signifying fruit production. Man, the primeval risk-taker and gene disperser, looked at far pastures and the promise thereof, where large prey and unattached women might lurk. These notions have occasioned lively discussions between my wife and me, but not a lot of disagreement with Orians and Heerwagen.

• • • • •

Aged nineteen and heading out into the world, I encountered the primal human habitat my first year at university. It felt like a religious experience akin to what the campus preacher and some of my friends described as being saved.

In the years before that time, tall hardwoods and filtered sunlight had circumscribed most of my days. Occasional forays into nearby, more open pine landscapes had awakened new and strong emotions. Before sunrise on a Saturday morning, something intangible would pull me from the dark hammock where we lived into the open pine barrens. Among the longleaf pines lay my best habitat at the time.

Bob Barsch, my roommate at A&M, lived in the central Texas "Hill Country" near the town of Brady. I first went there with him that fall of 1960. As soon as the car we shared with other Aggies going home for a long weekend climbed onto the Edwards Plateau west of Georgetown, something clicked in the hinterlands of my brain. It was not so much the expansion of view, which already had tracked us from College Station across the farm-scabbed prairie I came to know as Blackland. Rather, here on the plateau the trees suddenly shortened and withdrew to isolated clumps, and stony ribs of earth peered through close-cropped grass. Signs of people shrank to limestone ribbons of gravel leading to secret prospects beyond distant tree groves. The cattle looked rangier, spookier; goats and sheep lent diversity to domesticity. Most visceral of all, white-tailed deer bounded occasionally between tree clumps or peered from within them, and a flock of wild turkeys flew across the asphalt ahead. The wild and the untethered began to dilute the tame and the owned.

The sign outside town said Brady, Heart of Texas, population somewhere around 5,000. The vacant wool and mohair barn by the railroad tracks announced the decline of sheep and goat ranching and the economy in general. Shovel met rock a short distance below the soil surface. The dry air gave relief from the sluggish atmosphere I'd inhabited to date.

With Bob I explored the surrounding countryside. Limestone bluffs with overhangs and great cracks reared above small valleys with translucent streams. In the shine of headlights at night, eyes of ringtail and raccoon flashed from cliff face and brush heap. In the daytime, deer and turkeys flitted close. Bob's father drove us to the San Saba River, where channel catfish sucked our bobbers under in the swirling eddies.

The dark forests of my childhood seemed so distant, so small. Here before me lay the purpose of dispersal. The grass stretched short and sparse. I had come home to the Pleistocene.

SIX Where the Short Grass Grows

Figure 6. Pronghorn antelope on shortgrass prairie. Photo by author.

Mike Rose claims the ability to envisage the natural human habitat on the basis of the last bone in the big toe. Mike lives with his wife, Cordelia, on a shortgrass mesa just north of the small town in which I live. From their yard the view sweeps 360 degrees to horizons of near and far mountains, giving the impression of standing on a platform at the edge of a huge, shallow bowl. The night darkness, punctuated only occasionally by artificial light, stretches pregnant with prospect in all directions.

The word *savanna* best describes the hillsides and valleys that drop from the edge of the Roses' mesa. Grasslands dotted with juniper, scrub oak, and pinyon pine roll away to the north, the west, and the south. The Mogollon Mountains rise to the east.

A photo composite attached to an inside wall of their house shows a landscape of Kenya in East Africa. "One reason we retired here," says Cordelia with the British precision of speech she and Mike share, "is that it feels like Kenya. It reminds us of the years we worked there."

Our town occupies that part of the American West inhabited by foreseers and prophets of the intuitive kind. You might therefore conjure up an image of Mike rattling toe bones in a small cup and casting them, like dice, onto a Navajo blanket, then examining them for clues—looking for magical insight into the natural human habitat. But I have not known Mike to call upon clairvoyance. He is an anatomist and student of skeletons. For years, twelve in East Africa, Mike studied the bones of primates—humans, apes, monkeys, and their relatives. He retired from his university position in 2005. In early 2006 he and Cordelia went away for several days to Duke University, to a symposium held in his honor, a testament to his standing among those who study such things.

"The distal bone in your great toe," Mike says, "shows anatomical features peculiar to long-distance striders." These features show people to be the only living primates built for walking. Other primates have toe bones that look different; their feet, and the bones in them, serve best for climbing trees, not walking upright. The gorilla, the orangutan, and our closest relative, the chimpanzee, all live in forests. The human foot evolved in grasslands.

"My friend Rick Potts," he said, "recently published a book about human ancestry. You may want to read it." Potts's book appealed to me right away because of the foundation of his knowledge: years spent digging in the Rift Valley. He knew the habitat first-hand. In this region of East Africa, says Potts, we find the oldest-known signs of humans and near-humans. Here lies the story of our ancestors.

Potts offered insight into a question that had long intrigued me. At what point did evolution propel us from the line of the ancestral "ape-man" onto the tangent that made us human? Potts has a definite opinion: it happened when we left the trees and strode erect across the grass, when the bones of our legs lengthened and those of our feet relaxed and arched away from the curling grasp of tree climbers. In that faraway time we became hominids—members of the human family.

How long ago did this happen? Three to four million years, maybe longer, say the ancient bones and tracks. Potts notes several early hominid finds: Meave Leakey's *Australopithecus anamensis* from northern Kenya, Donald Johanson's well-publicized "Lucy" from Ethiopia, and Mary Leakey's upright strider's tracks preserved in volcanic ash in Tanzania. Some scientists see in these and similarly aged bones a residual dependence on trees, he said. This is not surprising, however, considering the typically slow, transitional nature of evolution.

Potts elaborates on what he and other anthropologists call the "savanna hypothesis," the notion that the drying out of East Africa's climate several million years ago, and the resulting change from forest to savanna, opened the way for an upright ape. Potts takes the idea further, arguing that not simply the increasingly arid environment but also repeated shifts in climate and habitats opened the way for a physically adaptable and problem-solving being—our ancestor. He puts it succinctly: "Adaptation to open terrain was the spark that initiated the human lineage."

The movement from forest to grassland gave access to a prized source of nutrition infrequently used before that. We call it meat. Our closest surviving relatives—chimpanzees, gorillas, orangutans—eat meat when they can get it but consume mostly leaves, shoots, fruits, and nuts. Our tree-climbing ancestors probably did the same. Dense forests provide little forage for large animals and yield only small amounts of meat. Out on the grass in that ancient time, however, fed an abundance of antelopes, giraffes, zebras, warthog relatives, and the predators that ate these herbivores.

Many students of human history think the first hominids got most of their meat by scavenging. Perhaps they learned to coopt prey from lions, hyenas, leopards, and some now-extinct carnivores. By eating meat they improved not only the quantity but also the quality of their food. Unlike plants, meat usually provides a well-balanced diet no matter what the source.

How did they take kills from great predators in the time before mastery of fire and spear? By waiting patiently in trees until the predators ate their fill, ganging up on lone lions or leopards, yelling loudly, throwing stones—we can only speculate. Desperate needs elicit desperate means.

The advantages of scavenging grew when these new savanna dwellers learned to fashion cutting instruments from stone. Anyone who has knapped flint or felt the edge of a prehistoric flake knows an effective knife can come from a few simple strikes of stone against stone. By contrast, human tooth and fingernail rip hide from carcass and meat from bone with clumsy inefficiency.

As long ago as two million years, processing carcasses with flakes of stone already had become tradition in our family tree. Flint scars on ancient bone bear testimony, say Potts and other excavators of animal remains. Swift butchering reduced risks from predators returning for another meal. Like coyotes scavenging after wolves, we teetered on the knife-edge between eating and being eaten.

Potts talks of analyzing several thousand fossilized animal bones unearthed from an Olduvai Gorge site about 1.8 million years old. The concentration of flint-scarred bones from several species made it clear that hominids had brought them from different habitats in the surrounding landscape. Gnaw marks of hyenas and other carnivores also scored some of the bones. Perhaps, Potts suggests, our ancestors assembled the bones, presumably with meat attached, and the carnivores cleaned up afterward.

Eventually our predecessors acquired the tools and skills for killing large animals, and hunting began to supplement scavenging. Probably the new capability gave rise to greater numbers of our ancient kin. With the new weapons they may have turned some of their previous predators into prey. In his book *In the Dust of Kilimanjaro*, conservationist David Western talks of present-day Maasai warriors in East Africa's Serengeti killing lions with spears and hyenas with clubs.

Rick Potts suggests that large animal abundance and diversity in Serengeti National Park today might resemble that encountered by our ancestors. Since Pleistocene times some species, such as saber-toothed cats and short-necked giraffes, have become extinct, and some new ones have arrived. But the climate, the vegetation, and the amount and chunk size of meat on the hoof may not have changed a lot. Can we, then, travel to the Serengeti today and see our natural habitat?

For more than a quarter century, Sam McNaughton of Syracuse University in New York has studied wild grazers on the Serengeti Plain. These animals—wildebeest, zebras, gazelles, buffalo, and others—contribute the preponderance of food available to meat-eaters. This is to be expected given that grass far outweighs other forage plants. Grass continues to fuel the great wildlife spectacle of East Africa today as it has for the past several million years.

McNaughton has documented an interesting phenomenon about the grazing animals: most prefer the short grass over the taller grass nearby. Both the migratory millions and the resident grazers usually prefer to feed in the sites where grass seldom grows higher than a few inches. They return to the same shortgrass swards year after year.

What underlies this seemingly illogical phenomenon of spots with less food attracting more animals? My range science professors at Texas A&M taught me that taller grass was better grass. Not so, says McNaughton. Grass kept short by grazing generally provides better nutrition, and the animals can taste the difference. Grazing eliminates most of the hard-to-digest stems and dead parts, and grazers fertilize the grass by urinating and defecating. In ecological terms, grazing speeds up nutrient cycling. The bottom line is that, most of the time, shortgrass terrain consistently supports more meat per acre than nearby places where the grass is taller.

McNaughton calls these favorite Serengeti feeding places "grazing lawns." Supporting bermuda grass and other low-growing species resistant to grazing, they contrast with the knee-high to shoulder-high grass in less grazed places. Grazing lawns are magnets for the vast herds captured on National Geographic film, and for the tourists who pay for the African experience. No doubt they were magnets as well for those ancient striders with the odd-shaped toe bones who found shorter grass not only easier to traverse than taller grass but also more productive of meat.

George Schaller studied lions in the Serengeti and notes another reason the great herds prefer short grass: tall grass hides lions. Because lions can outrun wildebeests and zebras for only short distances, they depend on a stalking or lie-in-wait strategy. That doesn't work when there's nothing for them to hide behind. So smart lions don't follow the migratory herds into the shortgrass plains; instead, they wait for them to venture into the taller grass and shrubbery along watercourses and in thicker woods. Smart wildebeests and zebras enter tall grass and thickets with great reluctance.

Logic suggests that ancient peoples, too, would have preferred those places with greater visibility and fewer stalking predators. To humans, stalkers epitomize the scary. They hide from sight, our most acute sense. They pounce before we have time to assess what to do.

Cougars hang out in the rough terrain near my town. People sometimes see these so-called mountain lions uncomfortably close. Their kills—commonly deer but also donkeys and barnyard sheep—show up near town. As a biologist, I know by the statistics that cougars seldom attack people. So, at least statistically, I risk little by hiking about in their habitat. But attacks by cougars on people do seem to be increasing. In the summer of 2008 near Silver City, an hour's drive from my house, a lion killed a man and ate part of him.

A few summers ago when out hiking near dusk, I heard the bleat of a deer fawn close at hand. Scrambling up a low ridge, I spotted a cougar with the fawn pinned to the ground, much as a house cat might pin a small rabbit. The cougar seemed to have the fawn's throat or nose in its mouth, and its ears were laid back. As I watched, the fawn gave one last kick with a hind leg, then lay still. Ignoring the mantra of letting nature take its course, I threw rocks. One landed near the cat, and it released the now-dead fawn and slithered into a nearby thicket. I retreated homeward in the gathering gloom, stopping often to peer behind.

Now on most hikes when the sun gets low, the shadows creep up, and thick cover hangs close at hand, I carry a long stick. The stoutest and lightest ones come from the dead flower stalks of the yucca-like plant called sotol, easy to find at our elevation. I tell myself the stick helps me balance on rubble-strewn slopes. But once the bushes fall behind and the visibility improves, I breathe easier and sometimes throw away the crutch.

Short grass: the most navigable terrain for our oddly shaped foot. The meat larder in that corner of the world where our kind took shape. The best protection from the skulking predators that watched our ancestors from five million years of stalking cover.

SEVEN Turf

Figure 7. Turf sport of the gentler sex. Drawing by Casey Landrum.

My first sense of chronic failure came on a bermuda grass field at the hands of a two-hundred-and-fifty-pound behemoth named Roy Harris. I already knew occasional failure, having grown up with the perfectionist tendencies of my father. He so seldom praised my brother and me that we got used to being not quite good enough.

I dealt with Daddy's tendency to criticize by doing well in conspicuous activities I knew he valued. Making good grades in school, completing work assignments, and showing courtesy to older folks topped his list. I also played a fair game of soft-ball in grade school, which I hoped he noticed. But I kept secret or avoided endeavors not likely to meet with his approval or standards of performance.

But it was hard to get by Roy. I got into the predicament by signing up for football my first year in high school without knowing much about it except the glory and the girls it attracted. The coach put inexperienced kids from the sticks, like me, on defense. You'll play man-over-middle, he said. I found myself facing off to Roy, the second-string center.

What I remember most about those long autumn afternoons on the practice field was the slamming together of shoulder pads, the crushing weight of Roy, and the smell of grass from two inches away every time the opposing fullback wanted to go up the middle. For three hours daily, turf in its physical and metaphorical dimensions governed my existence. I got stronger, but so did Roy.

The coach told me what to do. "Come to grips with Roy. Get low. Push harder." To me the best option seemed avoidance, but apparently that wasn't in the rule book. This was my introduction to the frustration of trying to succeed by criteria developed by someone with another agenda. Obedience, not thinking, seemed the measure of success on the football field.

At the end of the season I dropped out of football, a heavy blow to my self-esteem. There went my visions of rushing across the goal line to the standing applause of hundreds and the screams of leaping cheerleaders. My father never passed judgment, though now I think he probably approved of my switching to basketball and track, sports more tolerant of individual idiosyncrasy.

• • • • •

Did competitive games arise to entertain people, or train them for survival? Given my tendency to look for similarities between people and other animals, I think both. Young mammals play, probably because they like it, and by doing so they practice survival skills. Kittens stalk and rush; puppies go for the throat and legs. Our games teach boys to bludgeon, whack, kick, throw, and skewer. Girls can do these things too, though they usually adopt more subtle tactics.

Not surprisingly, nearly all our outdoor games seem to have developed on bare ground or turf. Our foot shape demanded it, and the heavily stocked pastures surrounding the settlements of civilization facilitated it. Given the other options available—tall grass, shrubland, thicket, forest, and marsh—outdoor sports show strong selection for short grass.

Turf games as we know them—that is to say, with written rules—seem to have evolved from the simple pastimes of illiterate peoples. Centuries of play on the grazing commons lie behind the modern versions of rugby, football, field hockey, baseball, cricket, soccer, tennis, bowling, croquet, golf, and others. Those who originally developed the games go unmentioned in sports history's halls of fame.

Rugby ranks among the earliest and most warlike sports of the sheep pasture. Written history dates rugbylike games back to at least the British Bronze Age, three to four thousand years ago. Eighteenth-century Celts called their version of the game "Caid," in reference to the bull scrotum that served as ball cover. Rugby Boys' School in England gave it the modern name and a formal set of rules in 1823.

American football is derived from rugby. Historians say it originated in the early 1800s as a distinctive game on the campuses of Princeton and Harvard, played by well-to-do young men looking for a way to vent their privileged energies. After the Civil War, formal rules took shape under the direction of Princeton players, the game acquired status as an intercollegiate sport, and the ball itself gained an official patent. Today, sophisticated body armor, rules designed to reduce injury, and sports medicine give fans the appearance of the brutality they want without unacceptable damage to players.

Croquet broke the rule that turf sports belonged to men. According to the Houston Croquet Association, the literate world learned about the game when a Scottish lady visiting the Continent in the mid-1800s saw French peasants at play. She took the rudimentary implements and rules back to the British Isles. There among the leisure class, croquet caught the fancy of newly liberated women. It became their sport. Men could join in, but only when properly chaperoned. The men soon introduced a version called "tight croquet" that allowed them to knock a lady's ball into the bushes and follow her after it.

Golf, now the favored game of the wealthy and the well positioned, probably evolved from a stick and ball game played by pastoral peoples in medieval Europe. The Scots claim to have invented it, but they probably only tweaked it a bit from French, German, Dutch, and other mainland versions, giving it the setting and spin to attract royalty. It became so popular in fifteenth-century Scotland that a succession of kings (James II, III, and IV) ordered the citizenry to stop playing because it interfered with the archery practice deemed necessary for national defense. In 1744 the first known rules for golf appeared on paper in Edinburgh. Today's faithful recognize the Royal and Ancient Golf Club, founded at St. Andrews, Scotland, in 1754, as the world's first such organization. Golf enthusiasts still go there to play.

• • • • •

My wife's stepfather, Rollie, lived sports. As a young man, Rollie as coach had led the Fenger High School football team in Chicago to unprecedented achievements. Once he showed me a book he'd written on football. Many of the pages had diagrams of little circles and crosses representing opposing players. Arrows showed the man-over-middle how to slide past Roy and make the tackle.

A television with a large screen dominated the living room of Rollie's retirement home in Tucson. From his favorite chair he advised players, cursing and fuming at errors in football, basketball, and baseball alike. He taped for later viewing the games he couldn't watch live.

Above all, he loved golf. He made the pilgrimage to St. Andrews after he retired. It seemed to consummate some lifetime longing, an achievement of the ultimate.

Once when he and his wife, Ceil, were still in their own house, my wife and I offered to help him with yard chores. He had us weed a sacred patch of crew-cut grass, about a hundred square feet—small to meet Ceil's water-saving ethic, I suspect. He pointed out a kind of low-growing spurge that had sneaked into his stand. I thought it added appeal, but he said no. An hour's hand work by Judy and me brought it once more to putting-green standards.

Experiencing Rollie's yard and those of countless others over the years has given me cause for reflection. Why the short, immaculate surfaces, the monocultural sameness fronting house after house, from scattered settlement in the Alaskan forest to small prairie town to large southern city? The uniform green grass and the whirring lawnmowers on summer Saturdays fit poorly with my concept of the meaning of life. Why the rule of ordinance and censure of neighbors to bring the maverick yard-owner with tall grass or vagrant thistle into line with the rest?

The power of short, green turf as symbol of status goes far back in our history, says Virginia Scott Jenkins in her book *The Lawn: A History of an American Obsession.* She traced its lineage to prerevolutionary times when newly arrived Euro-Americans still claimed allegiance to the motherland. Anthropologist Rick Potts might put its roots much, much farther back, to the grazing lawns of ancient Africa, the original habitat of us all.

• • • • •

To me, one of the great disappointments of American history is the failure of those who came to dominate the landscape after Columbus to improve on their heritage of land use. They could have learned much from those already here and built a different vision of our place among the other species. But they didn't.

At the end of the Revolutionary War we latter-day Americans, with celebrations in the streets and a flourish of signatures, sent the British packing. We proclaimed our independence from our metaphorical fathers in the motherland. Yet although we didn't want them meddling in our affairs, we continued to ask them how we should live, and what should front our houses.

Like wealthy Europeans of the day, Americans of prominence advertised their status by manicuring their estates, copying, as best they could, the Old World terrain of contest and leisure. The ink had scarcely dried on the peace treaty ending the Revolution when Thomas Jefferson hurried overseas to view the habitats of nobility as model for the grounds of Monticello. What he saw had a peculiar history, one we Americans would be destined to repeat. It had to do with exploitation, wealth, and the accumulation of status.

Great Britain's estates of the day derived from her wealth. That wealth came from a unique combination of factors. Among them were expansion of trade around the globe, milking of money from overseas colonies, and enhancement of production at home.

Imperialism and international trade worked hand in hand. Britain's ships more or less ruled the waves, I learned in grade school, and the sun never set on her empire. Trading always has been a good bet for making money, and the slave trade turned out to be particularly lucrative for the British. Skimming from the labors of New World peasants did not cease simply because an ocean separated the laborers from the coffers.

Wealthy Britons in collusion with Parliament removed most of their own peasants from the land under a series of laws called the Enclosure Acts. Several thousand of these place-specific rulings took shape in the century following 1750. Peasants lost their grazing commons, their small farms, their rights to scavenge grain from fields of the wealthy, and thus their ability to survive in the countryside. They then swarmed into the grime-coated cities of Charles Dickens and the Industrial Revolution. British poet Oliver Goldsmith, in his epic poem "The Deserted Village," lamented the passing of the long pastoral era in England's countryside, the exodus from "Sweet Auburn! Loveliest village of the plain / Where health and plenty cheered the laboring swain."

Outputs from factory and farm mushroomed following land enclosure. Peasants made good factory workers. The estates of the wealthy, cleared of their peasantry and cramped inholdings, expanded to sweeping panoramas unobstructed by chicken coop and pigsty. Thomas Jefferson, visiting American, champion of the yeoman farmer, exulted in the view. He took the image home to Monticello.

Virginia Jenkins tells how other wealthy Americans of the day, like Jefferson, copied European custom. George Washington—who never traveled overseas—hired English landscape gardeners to shape the grounds of Mount Vernon. Servants of William Hamilton trimmed and scythed his estate to British perfection. John Penn of Pennsylvania kept the grass on his estate short by employing a picturesque shepherd with sheep and herding dog. Indeed, sheep grazed most New England estates in that time before mechanical lawn mowers.

Not only did the early estate designs and gardeners come from Europe, so did the grass. Jefferson and most others in the cool New England climate favored Kentucky bluegrass—which came from Europe, not Kentucky. Southern gentlemen eventually turned to heat-adapted grasses such as bermuda—originally from Africa. The preference for Old World grasses resulted less from nostalgia than from necessity. The native bunchgrasses of the American seaboard did not form a turf under close grazing; rather, they withered and died.

Grass came to advertise differences in economic class in America much as it had in Europe, says Virginia Jenkins. During the time Jefferson watched his slaves trim and polish his Monticello estate, and for many decades afterward, less wealthy Americans did not spend a lot of time grooming the landscapes surrounding their homes. Instead they measured their status by the frequency of food on the table. Packed clay or sand sufficed for their yards.

In the 1940s, both my sets of grandparents lived in houses surrounded by clean-swept yards of sand. My brother and I asked for spoons when we visited, so we could dig holes in the smooth, hard surface and shoot marbles at them from a distance—a kind of golf without clubs. Not least, the high visibility made it easy for us to spot copperheads and rattlesnakes, and the lack of grass discouraged ticks and chiggers.

But my parents seemed to like grass, and it covered our yard—much to my dismay. You couldn't dig in it with a spoon or shoot marbles across it. It had to be mowed. Bermuda grass gave me red welts when I rolled shirtless in it.

Were Daddy and Mama striving for the status Virginia Jenkins described as coming with grass-covered lawns? I cannot imagine they were, at least consciously. Daddy often criticized status-seeking behavior, and

Mama tended to go along with Daddy. But, as Jenkins notes, the marketers of yard status in America worked in devious and subtle ways.

The beginning of the grass lawn aesthetic for the middle class paralleled the rise in American affluence following post–Civil War Reconstruction. As the nineteenth century neared its end, the American economy boomed, particularly in the North. The flaunting of wealth and leisure by the more fortunate gave name to the Gay Nineties.

As usual, the privileged set the standard. For aspiring men, the golf course served as reference point; for women, the front lawn advertised standing. Such gender differences in habitat preference may seem inappropriate in this liberated age, when many women golf and men often inhabit yards more than do their female partners. But Jenkins and I both suspect that, deep down and on average, men prefer the far prospect of golfing and women the near refuge of the yard.

The first permanent American golf course took shape in 1888, in a cow pasture near Ardsley, New York. Not surprisingly, a Scotsman—John Reid—laid it out. Four years later, Rhode Island and Long Island had a course each, and Chicago had two. The trend spiraled upward from there, inspired by Americans wealthy enough to tour abroad and bring home the British model. Golf became, and remains, the sporting icon of the wealthy.

Golf offers magnificent prospect. The course itself mimics the ancient grazing lawn, the sward that in the deep recesses of the male brain promises meat with minimum risk. Trees stand far enough apart to allow one to see the archetypal predator at a distance and yet climb quickly to safety if necessary. Pools of water await the thirsty forager. Pathways curve away to tantalizing mysteries beyond the next hill, around the next copse. For men, the crowning attraction may lie in the action: only in this sacred place can one hurl projectiles so far and at such fantastic speed.

Front lawns as desirable adornments of middle- and lower-class homesteads arose from the Garden Club Movement of the 1890s, says Jenkins. As the twentieth century gained momentum, such clubs spread across the country under the sponsorship of well-to-do women with money and leisure time. Campaigns for community beautification decried the bare, littered, and unkempt landscapes that commonly surrounded homesteads of the poor. In my county, the Pine Burr Garden Club of Jasper set the local lawn standard.

By the time of my birth, according to Jenkins, the Garden Club front lawn symbolized status for up-and-coming Americans. Patterned on European fashion filtered through the higher classes in America, it had crept to the front of most middle-class households for all to judge. Looking back, I can see how my father and mother might have caught lawn fever by gentle contagion from their friends and neighbors. But my grandparents clung to the bare and the broom-swept.

Once it became evident that wealthy people wanted turf and were intent on converting the less wealthy, the U.S. Department of Agriculture (USDA) took on the task Thomas Jefferson had started a century before. Federal agents began to look about for appropriate species. In 1897 the USDA's new Division of Agrostology took official charge. Its scientists found, as had Jefferson, that all the good lawn grasses had to be imported.

Wasn't American grass good enough? Usually not, as it turned out. The low, creeping grasses most suitable for lawns had evolved under thousands of years of heavy grazing. Good turfgrasses had developed alongside the great herds, wild and domestic. So not unexpectedly, the best turfs came from Africa, Europe, and Asia.

James G. Beard, world authority on turfgrasses and their history, notes the historical relation between domestic grazers and turf in the Old World. Sheep and other livestock maintained playing fields and lawns at least as far back as biblical times in Europe and Asia, says Beard in his book *Turfgrass: Science and Culture.* He quotes instructions written by one John Rea in 1665 for digging up and transplanting sod from the grazing commons of the lower class to the leisure lawns of the upper: "The best turfs for this purpose are had in the most hungry Common . . . where the grass is thick and short."

What about the bison, the perceptive American patriot might ask? Weren't they abundant out on the Great Plains? If so, did not turfgrasses develop under their grazing? The answer is yes. But turf researchers and producers have shown little interest in them until recently. We will revisit buffalograss later on.

The "thick and short" grasses brought from around the world formed the basis for an American sod industry that now rivals the economies of small nations. The U.S. Environmental Protection Agency estimates that turf in America today, three-fourths of it on lawns, covers an area the size of Pennsylvania. Half of all American households have lawns. The lawn tenders—be they men, women, or teenagers—spend exorbitant hours mowing, weeding, and poisoning to keep their patches of habitat short, smooth, and uniform. Many spend yet more hours displaying themselves upon the habitat.

• • • • •

Under the backyard elm we prepare for a barbecue. Guests arrive, share ritual embraces and handshakes, repeat tribal salutations. Friends shed sandals to wriggle their toes in the new-mown grass. A hazy sun hangs ready to drop behind a skyline of human design. The agony of another work week at the office subsides. Children play. Fat sizzles over red coals.

As twilight thickens we draw together into a flickering knot. City sounds subside to a low rumble. Age-old juices of fermentation flow freely, dissolve the present. The close noises of motor and amplified music metastasize into the distant roar of lion and grunt of wildebeest. We drift away to a far savanna and a time long past. We stoke the fire. We are home.

• • • • •

One spring morning in 2002 I left Eugene, Oregon, and followed Interstate 5 north. Fields of green lay to either side, dotted here and there with sheep and cattle. To the right, abrupt hills drifted in and out of low clouds to flank the eastern edge of the Willamette Valley. My son Jed drove the car. He had arranged for us to meet a Mr. Ray Doubrava.

Ray managed the Oregon Turf and Tree Farms near Hubbard, about fifteen miles south of Portland. After an hour's drive, we pulled into his driveway. Nearby loomed large sheds guarded by big-wheeled trucks, and beyond lay a perfectly flat field striped irregularly with brown and green. Ray came out to greet us. "Our turf farm," he said with a sweep of his hand. "Three hundred and fifty acres."

We strolled among sheds and machines as Ray explained the operation. "We do baseball and soccer fields, commercial and private landscapes, lawns. We specialize in perennial ryegrass—perfect for this climate and green year round. Our seed supply—virtually weed free—comes from a dependable producer. Oh yes, those pastures you saw on the way up here probably were annual rye, related to what we grow. Farmers around here sow annual rye for their livestock."

My James Beard turf manual calls perennial rye *Lolium perenne.* Originally from temperate regions of Asia and North Africa, Beard says, it may be one of the first grasses cultivated by humans. We may have eaten its seeds long before we played on it.

Ray showed us his tractors with their great banks of reel mowers, blades freshly hand-sharpened to a razor edge. I thought about the status one of these would have given me as a kid mowing yards. "We keep our employees," said Ray. "They know their work. We have the ideal American small business."

He led us to the green fields. "Here's how we produce sod. First we plant the seeds. Then we lay out this biodegradable mesh over the seeds. In this way, when the grass grows through the mesh, it holds together for harvest and transport. Just makes it easier with ryegrass—some kinds of turf hold together well enough without reinforcement.

"Once the grass starts growing," he continued, "we mow several times a week. Makes the grass grow flat and thick. If you don't mow, it gets tall and spindly.

"At harvest time, we shave off the top inch or so of soil, including the mesh and the imbedded grass. Then, with this other machine"—he pointed—"we roll the long strips up, like a carpet, and load them onto these trucks. Sod's heavy; that's why the big tires. When we reach the job site we unroll the sod, piece together the strips, and you've got an instant lawn."

He swept his hand out toward the fields. "We do take away a thin layer of soil every time we harvest. You'd think we would eventually run out of soil. But it's deep here. We're on the old Willamette floodplain. They say the soil was laid down way back in the Pleistocene when huge ice dams periodically broke upriver and the muddy water spread out and dropped its load of silt. We're standing on soil a few hundred feet deep."

Thomas Jefferson would have appreciated Ray's business. One phone call and the pastures fronting Monticello, laboriously seeded and nurtured by slaves, could have taken shape in a day. The U.S. Golf Association and the City Beautiful Movement a hundred years past would have applauded Ray as well. And today, athletes, sports fans, and sellers of mowers and grass fertilizers appreciate his work. Though purveyors of herbicides might wish his grass had a few more weeds.

Even the working wives of Portland must be glad for Ray and his crew. In front of the farm's small office we saw house-husbands picking up small mats of Ray's sod at twenty-two cents a square foot.

• • • • •

The attractive dark green edition of James Beard's manual that I found in a used bookstore hints at the substantial costs of turfgrass research and development. For more than a hundred years now, the U.S. Department of Agriculture in collaboration with American universities—including Beard's own institutions, Michigan State and Texas A&M—have worked to develop grass varieties resistant to close trimming, foot traffic, pests, weeds, and weed killers. The favorite grass species have proliferated by artificial selection into hundreds of varieties, or "cultivars." Growers now can provide cultivars to suit every imaginable soil, climate, pest regime, and use.

The U.S. Golf Association helped initiate these federal programs, says Virginia Jenkins. Perhaps that's as it should be, since the wealthy benefit disproportionately.

The 2001 book *Redesigning the American Lawn: A Search for Environmental Harmony,* by F. Herbert Bormann and others, estimates the annual cost of lawn care products and equipment in America at $8.5 billion. Costs of turf and lawn maintenance adds another $30 billion. For those with lawns, that totals at least $200 per person per year, and perhaps $1,000 per household. All to decorate a piece of ground that could, with the same input of energy, fertilizer, and water judiciously applied to edible plants, more than feed the average household.

My grandfather Corbett and grandmother Fannie would have wondered at the cost. Their yard expenses probably reached no more than a few dollars per year. Valuing food more than Garden Club status, they put most of their effort into the vegetable garden.

Shrewd advertising by the lawn-care industry coupled with our deep-seated attraction to short grass and status have prompted many of us to surround ourselves with exaggerated versions of the natural human habitat. Those with perfect grass—a single species, one shade of green, uniform height, and a large expanse that must be mowed frequently—inspire the rest of us to strive for such perfection. The symbols of wealth and prestige continue to drive us as they have for so many thousands of years.

• • • • •

Once a year the main street of my town bustles with people in baseball caps and big hats come to play cowboy golf. Just off the main street to the west stretches an irrigated field of domestic grass. Cropped short by Angus cattle, pocked with cloven tracks, and dotted with brown patties, it awaits the event. Ordinary people carrying golf clubs, balls, and tees exit their vehicles—mostly pickup trucks—and pass through the pasture fence. Entering upon the original home of golf, the grazing sward, they take back for one festive day the game stolen from their ancestors by the privileged class.

EIGHT Grass and Grazers: An Ecological Primer

Figure 8. American icon on the Armendaris Ranch, New Mexico. Photo by author.

Very soon after becoming third president of the United States, Thomas Jefferson determined to consummate a long-held ambition. He wanted to explore the interior of the American continent and map a travel route to the Pacific Ocean. But encumbered by the presidency and advancing age, he could not think of doing it himself; he would have to depend on someone else. So he sent word to a young friend who carried the curious name Meriwether Lewis.

In March 1801, Lewis hastened to Washington from where he lived in Pittsburgh. For the next two years he lived in the White House as Jefferson's personal secretary. The two pored over maps, schemed, and planned. Jefferson lobbied Congress for money to finance an expedition. He sent Lewis to Philadelphia for a crash course in science and instructed him to pick a second in command.

By winter of 1803–4 Lewis, his new partner William Clark, and their young crew had set up camp near St. Louis, the western outpost of European-style civilization in the heartland of the continent. The next spring, with the arrival of warmer weather, they crossed the Mississippi River, entered the mouth of the Missouri, and proceeded upriver under oar and sail. Thus commenced an epic journey that would take them across the middle of the continent, where few Europeans except French traders had ventured. Soon they would encounter unimagined prospect: a vast and wild landscape of people, animals, and grass unrivaled among the grazing pastures of the world.

As spring of 1804 merged into summer, prairie with grass high as a man's head drifted by on the land that rose above the river. Near present-day Columbia, Missouri, on June 6, the travelers saw their first bison sign, though it would be a long time before they reached the great herds. Twenty-two days and two hundred river miles passed before they even saw a bison, near what is now Kansas City. Here the river veered north.

In late August and early September they made their way past present-day Nebraska and Iowa and into South Dakota. The grass began to shorten. Only after they had moved deep into South Dakota did the larger bison herds and the people and animals who lived with the bison begin to fill the pages of their journals. They gloried in what those who came later would call the Great Plains. On this stage there arose concepts of ecology that would change the way people worldwide thought about and practiced pastoralism.

• • • • •

A hundred years after Lewis and Clark passed up the stretch of the Missouri that separates Nebraska and Iowa, a young man born in Villisca, Iowa, fifty miles east of their route, entered university in Lincoln, Nebraska, fifty miles west of their route. Prairie raised, he showed an unusual interest in grass. It would become his lifelong passion. In time, he would leave a legacy of grassland study unrivaled then or since. As a student at Texas A&M, I used one of his books, *Plant Ecology.* His name was John Weaver.

Weaver happened onto the stage of ecology at a momentous place and time. At the University of Nebraska, he came under the influence of the famous Charles Bessey, considered by many the father of modern botany. Frederic Clements, himself a student and protégé of Bessey, supervised Weaver in his undergraduate work at Nebraska, and later in his Ph.D. program at the University of Minnesota. Henry Cowles, eminent ecologist at the University of Chicago, added froth to the midwestern commotion of new ideas in community ecology that soon would spread around the world.

The most influential of the ideas that emerged from this small group came to be called the theory of ecological succession. We touched on it earlier. It holds that natural communities, when freed from unnatural disturbance, rise with a purposeful intensity through a succession of so-called seral stages to a final and stable state called climax. Succession theory revealed itself to Clements through the dynamics of prairie, and to Cowles through his study of vegetation on Lake Michigan's Indiana Dunes.

As a student, John Weaver accepted these great men as minor gods, their ideas as gospel. The notion of climax colored his perceptions thereafter of right and wrong in the natural world, as it has the perceptions of so many since. Nevertheless, some scientists then, and more in recent years, challenged succession theory as a semireligious parable unsupported by science. Later we will talk more of the pros and cons of succession theory, but for now rest assured that John Weaver's observations of the grasses themselves remain unchallenged.

In 1915, Weaver returned from Minnesota to take a professorship at the University of Nebraska. He never left. There in the fertile prairie soils, in the American breadbasket where farmers still claim to feed the world, John Weaver tried to ignore the burgeoning fields of corn, the North American grass turned domestic. Instead he sought out the diminishing parcels of wild grasses. True to the admonitions of Bessey, Clements, and Cowles, he returned to the land, year after year, to catalog, measure, dig, observe, and photograph.

He wrote about grass, his grass, his world. His 1954 book *North American Prairie* summed his accumulated knowledge of the subject, gained through "more than forty summers" out with the grass. He called himself fortunate to live in that portion of the prairie that had "resisted civilization" longest. He explained the origins of prairie soils. He revealed the grasses' roots, reaching six, eight, and ten feet deep to build the great reserve of fertility on which so many around the world have feasted for so long. He described the forbs, or broad-leafed plants, that mingled with the grasses. He showed the changing faces of the prairie with changes in weather, soils, topographic relief, and disturbance.

No one can teach, by book or in a classroom, the love of grass that consumed Weaver. Like a mellowed marriage, that kind of affection comes only with time spent together. It grows from naming, touching, and smelling your fellow traveler under sun, wind, and rain. It requires immersion in habitat.

Grass, like other faces of nature or like a person with depth, reveals its greatest charm in its complexity. But loving anybody or anything complex requires a period of learning, and learning requires that we name and classify. Weaver found useful ways to classify.

How tall is the grass? Because we stand erect, stride on flat feet, and explore our surroundings by looking, grass height means a lot to us. Weaver talked of three height categories: tall, mid-height, and short. These have become standard ways of describing grasslands, grass stands, and even grass species. Tallgrass species range from about three to ten feet tall (often too high to see over), midgrasses one to three feet (tall enough to hide a sneaky predator or big snake), and shortgrasses less than a foot (the shorter the better for the human strider). You'll hear ecologists and cowboys alike speak of tallgrass prairie, mixed-grass prairie (which reaches mid-height on average but usually contains a mix of short, mid-height, and sometimes tall species), and shortgrass prairie (which some scientists call "steppe").

Is it cool season or warm season? For millennia, pastoralists in temperate or northerly latitudes undoubtedly have noticed that some grasses green up and set seeds in cool months but others do so only in summer. More recently, plant physiologists discovered the reason: the two types have different chemistries of photosynthesis. As the type names suggest, warm-season species tend to dominate at lower latitudes, cool-season ones farther north or higher in the mountains. Grassland ecologists and cattlemen know that in temperate zones a good mix of the two types carries large grazers year round better than does either alone. In climates with distinctive seasons, lawns stay green longer over the year with a mix of types.

Does it grow in clumps or form a turf? The taller grasses tend to rise from well-spaced clumps, while short grasses often spread like a carpet, especially if they are grazed or mowed. Some species can do either, depending on rainfall or grazing intensities. Some shortgrasses, like the European species imported by American sod growers, work well as playing fields and lawns.

Finally, is it annual or perennial? Perennial grasses dominated the grasslands that Lewis and Clark saw in the summer and fall of 1804, that John Weaver studied over a century later, and that you see today most of the way from New Mexico to Montana if you drive Interstate 25. The individual plants live ten, twenty, or more years and send down long roots that give the dark cast and lasting fertility to true prairie soils. Annual grasses live a year or less, overwinter as seeds, and colonize waste places and grain fields. They feed few free-ranging cattle but most of the world's people.

John Weaver showed why grass height, season of growth, and longevity made a difference in America's heartland. Tall perennials maturing in summer sustained the hay production that kept most of his study sites free from the plow. They also produced more tonnage per acre than midgrasses or shortgrasses. But shortgrasses weathered drought and grazing better than tallgrasses and kept their nutritional value better through the winter. Annual grasses found little opportunity to grow at all except in areas laid bare by drought, grazing, or the farmer's intent. Eventually, however, nearly all Weaver's beloved native pastures would give way to annuals—corn, wheat, barley, sorghum, and other high-value crops.

Weaver plowed new scientific ground in the prairie. He studied roots of perennial grasses—the secret of prairie fertility, of grass resilience to drought, of grass persistence under heavy grazing. Here the grasses store their reserves in winter and retreat from grazers, heat, and drought in summer. Weaver laboriously washed roots free of soil, then sketched and photographed them. His picture of a six-inch-tall buffalograss wedge with six-foot roots looks like Rip van Winkle with a crew cut.

Tallgrasses and most crops, unless irrigated, require plentiful rain, on the order of thirty inches or more per year. In America's midsection, rainfall, food production per acre, and carrying capacity for livestock diminish as one travels west. Precipitation tells why the eastern prairie region, with thirty to forty inches of rain, produces most of the grain and livestock, and why native tallgrass prairie occupies less than 1 percent of the land it used to. It also explains why, as you take Interstate 70 west from forty-inch-rainfall Kansas City to where rainfall drops to less than twenty inches, native grass shrinks in height but expands to the horizon. In these drier parts of the Great Plains, where ranch houses and range cattle stand far apart, farming thrives only with irrigation.

John Weaver's studies spanned the Great Depression and the coincident Great Drought of the 1930s. As the drought deepened, Weaver saw the taller grasses die or retreat to wet sites. Tallgrasses had too much leaf surface and too little root storage to withstand long periods without water. In the late 1930s and early 1940s, he and a colleague, F. W. Albertson, conducted a detailed analysis of a site in northern Kansas. As his beloved tallgrasses—big bluestem, little bluestem, Indiangrass—withered away, the soil began to move. Thus the native prairie began to add its soil to the Dust Bowl clouds that lifted from the endless plowed ground that never should have been plowed. Weaver mourned.

But as the tallgrasses retreated, a curious phenomenon showed itself: the shortgrasses—buffalograss and blue grama—began to expand. Released from the intolerable shade of the taller species, and needing less water, they spread rapidly into the bare spaces. The lowly shortgrasses, denigrated as seral or successional by Weaver and other climax theorists, ended up covering and holding the prairie soil.

The study's results appeared in a 1946 issue of *Ecological Monographs* under the title "Reduction of Ungrazed Prairie to Short Grass as a Result of Drought and Dust."

Like drought, heavy grazing seemed also the handiwork of the devil. When domesticated grazing animals were introduced, said Weaver, the prairie began to "degenerate." Tallgrasses gave way to shorter ones, and weeds began to invade. Extent of retreat from climax paralleled intensity and duration of grazing. Grazing by wild animals before the settlement of the prairie had little effect on the grass, he said, allowing the climax to persist.

But wait, cried Floyd Larson of the U.S. Soil Conservation Service from the back of the lecture room. What about bison? In 1940 he had written a rebuttal to the climax mania sweeping the country, an article published in the journal *Ecology* and titled "The Role of the Bison in Maintaining the Shortgrass Plains." He took direct aim at Clements, Weaver, and other believers:

> Certain plant ecologists refuse to give the short grass plains the status of a true plant climax, holding that this plant community is a disturbance or disclimax brought about by the coming of man and domesticated grazing animals and that the true climax is the mixed grass prairie, which would reappear if pristine conditions were restored. This paper attempts to show that the Great Plains was formerly grazed heavily by bison and other wild animals and that this primitive grazing held the western plains in a short grass stage which cannot be called a disturbance climax because this animal life was natural to the biome.

Who was right? Larson, and later others, showed that before European entry bison indeed had heavily grazed the grasses over wide regions of the Great Plains, probably for centuries. The result was domination by buffalograss, blue grama, and other shortgrasses highly resistant to grazing. Did not the astute Weaver know about this and appreciate its relevance to climax theory?

Lewis and Clark provided part of the answer. When in 1804 they transected the tallgrass country along the Nebraska-Iowa border, where Weaver later would develop his notions of climax, they found the prairies remarkably free of bison. Jim Shaw and Martin Lee of Oklahoma State University got the same impression when they analyzed journals and diaries of explorers and naturalists who traveled across the Great Plains in the half century after Lewis and Clark: few bison lived in the tallgrass regions. So perhaps Weaver and Larson both were right, each from his own perspective.

If this kind of conclusion leaves you with a sense of unfinished business, you are not alone. Why didn't Weaver's main stomping grounds—the tallgrass prairies—have many bison? Was that in fact "natural"? In 1997 Shaw and Lee published an article in *Plains Anthropologist* that suggested it wasn't. Although they agreed with Weaver's contention that few bison occupied the tallgrass prairie in the 1800s, that didn't, they said, make ecological sense. They knew that some managers running bison in the late twentieth century in tallgrass country had found that it could support very high bison densities. So why not in the nineteenth century? As befits good scientists, they qualified their answer as hypothetical: aboriginal horsemen living partly on agriculture in this high-rainfall region had already killed most of the local bison by the time European travelers came along.

Indeed, Lewis and Clark offered a relevant commentary as they passed through tallgrass country from present-day Kansas City to Sioux City, Iowa. They noted that most of the native villages, located in permanent sites and sustained largely by corn, contained few able-bodied men. Those men had ridden westward toward shorter grass in search of bison.

• • • • •

Those who study the origins of grasses and grazers concur that the two shaped each other over the course of fifty million years or so of coevolution. During this time, unimaginably long by the biblical calendar, grazers gradually acquired teeth that could efficiently crop and grind the coarse fiber of grass and digestive systems that could break down the cellulose-rich cell walls and absorb the nutritious parts. Grasses kept a step ahead, finding ways to avoid death from grazing and trampling. Like chicken and egg, neither grass nor grazer came first. Time, sunlight, and the patient ways of evolution gave them form together.

Different grasses, like different people, tend to specialize in different things. Some specialize in the interception of sunlight, some in the capture of water or nutrients, and some in the tolerance of cold weather. Eons of animals grazing at different intensities under different sunlight, water, temperature, and soil regimes have given different grass species different ways to avoid death. Few are simultaneously good at all things: that would take contradictory strategies or simply too much energy.

The different grassland animals likewise favor different kinds and combinations of grasses. Large grazers such as bison and cattle have the most catholic tastes, doing reasonably well whether grasses are tall or short, cool season or warm season, perennial or annual, and responding mainly to grass poundage per acre. Not so the small animals that constitute most of the vertebrate species. These—the birds, the rodents and rabbits, the smaller carnivores—respond to the physical structure of the sward. To them, grass height makes a lot of difference.

To appreciate these nuances of grasslands requires more than a casual glance from a Great Plains highway. It calls for learning the names and qualities of a few of the grasses visible from highway rest stops and some of the animals near the road in between. Otherwise the prairie and plain that take up where crops drop out as you drive west in springtime along Interstates 40, 70, or 90 may look little different from the monocultures of corn, wheat, or sorghum you passed the day before. Like people of alien races, grasses and road kills tend to all look the same until you get to know them.

Why bother to learn anything about such boring landscapes in the first place? In the modern world of flash and color, who takes the time to slow down and wonder about the web of life that constitutes a grassland? So far away seem the sensitivities of John Weaver: "One glories in its beauty, its diversity, and the ever changing patterns of its floral arrangements. But he is awed by its immensity, its complexity, and the seeming impossibility of understanding and describing it. After certain principles and facts become clear, however, one comes not only to know and understand the grasslands but to delight in them and love them."

• • • • •

As Lewis and Clark in 1804 passed from the mix of tallgrass and timbered terrain in Missouri and Iowa into northeastern Nebraska and South Dakota, they entered a more open country with shorter grass. They encountered a succession of animals new to them—pronghorn antelope, mule deer, bighorn sheep, white-tailed jackrabbit, coyote, swift fox, black-tailed prairie dog, magpie, and others. The species they encountered reflected the iron hand of grass height. They themselves, though accustomed to the wet and the close-wooded, voiced time and again spontaneous delight in the semiarid and the far-reaching. Had they, like the species they encountered, found their proper habitat?

NINE Bison Plains and Prairie Dogs

Figure 9. Bison hunting scene by George Catlin. Courtesy Yale Collection of Western Americana, Beinecke Rare Book and Manuscript Library.

Apart from its intended mission, the Lewis and Clark expedition turned out to be an interesting experiment in landscape appreciation. This was a group of young men raised Euro-American style in the eastern woodlands of late-eighteenth-century America. How did they adapt to a life of hunting and foraging in a vast interior grassland peopled by aboriginal tribes? What qualities of the landscapes they traversed appealed to them as they made this transition? Thomas Jefferson's instruction that they record in detail their observations and impressions resulted in an evocative and informative written legacy of tame dispersers gone feral.

Both Lewis and Clark hailed from northern Virginia, a center of American social and political life then as now. They grew to manhood in an agrarian-industrial economy with its attendant ways of valuing land. Young as captains of major expeditions go these days—Lewis in his late twenties and Clark his early thirties—both had learned military logistics in their early twenties. That experience explains in part the success of the expedition. They knew that, despite detailed planning and outfitting, they would spend most of their journey across the continent living off the land and by the good graces of the people already there.

Consulting a modern map lets us follow them as they left their winter camp where the Missouri River meets the Mississippi on May 14, 1804. The first leg of their journey up the Missouri—across the present-day state of Missouri west to Kansas City—took one and a half months. They fell into the routine of moving most of the expedition's approximately forty members upstream in boats—a fifty-five-foot keelboat rigged with sail and two pirogues propelled by oars. The expedition's hunters with a couple of horses followed them on land, scouring the nearby woods and prairies for food.

Nicholas Biddle's *Journals of the Expedition*, published in 1814, combined entries from the individual journals of Lewis and Clark. It took final shape under the editorial scrutiny of William Clark after Lewis died mysteriously in 1809. Biddle boiled the journals down to the essentials, with minimum alteration of phraseology. In reading his two volumes, one is struck with the changes in how the two captains responded emotionally to landscapes as they moved upriver.

Between St. Louis and Kansas City, both men showed mostly agrarian appreciation for the terrain. After all, their sponsor, Jefferson, viewed the ideal human condition as agricultural. They called favorable landscapes "fertile," "rich," and a couple of times, "pleasant." Utilitarian terms described both forest and prairie, but only the prairies were "pleasant." The mosquitoes, ticks, and humidity they described may have dampened their ardor for the mostly low-lying woods transected along this stretch of the Missouri. Nothing appeared "beautiful" until they neared the mouth of the Kansas River, near present-day Kansas City.

Here the Missouri River and their course along it veered northward. From there to the mouth of the Platte River—present-day Omaha—the aesthetic tenor of their landscape appraisals rose a notch. They saw an approximately equal number of "rich" or "fertile" places and those—invariably prairie—that were "fine," "pleasing," or "beautiful." Was the landscape improving in beauty, or was the dutiful farmboy falling by the wayside?

Beyond Omaha, "rich and fertile" lowlands almost disappeared from their entries. Timbered country withdrew to the streamsides and called forth little commentary. The glory of plain and prairie began to captivate the two captains, who waxed eloquent about the stretches of grassland: "high and handsome," "extensive and delightful," and frequently "beautiful." Wonderful "prospects" greeted them with increasing frequency from hill and bluff near the river.

Unedited versions of the expedition's journals published in the late 1980s by Gary Moulton and the University of Nebraska make the mounting excitement more tangible. The journalists' exact words leave little doubt of rising spirits as they moved deeper into the grasslands. Clark, less tutored than Lewis in proper English, exclaimed from an eminence where the Missouri first impinged upon what is present-day southeastern Nebraska: "I had an extensive view of the Serounding Plains, which afforded one of the most pleasing prospects I ever beheld, under me a Butifull River of Clear water of about 80 yards wide Meandering thro: a leavel and extensive Meadow, as far as I could See, the prospect much enlivened by the fine Trees and Shrubs."

At this place, where the Nemaha River enters the Missouri, the men also found grapes "nearly ripe," "plums Crab apls and a wild cherry," "Elk & Buffalo." A fruitful prospect indeed for the crew, now obligate hunters and gatherers.

On September 7, 1804, Lewis and Clark ambled away from their riverside camp to investigate a small "round mounting" they saw a few miles south of the river. At this landmark Clark described an animal new to them but long the companion of peoples and other animals in the American heartland: "Capt Lewis & my Self walked up, to the top which forms a Cone and is about 70 feet higher than the high lands around it, the Bass is about 300 foot in descending this Cupola, discovered a Village of Small animals that burrow in the grown (those animals are Called by the French Pitite Chien). Killed one & Cought one a live by poreing a great quantity of water in his hole."

By this cryptic entry Clark announced the "official" discovery of the black-tailed prairie dog. In succeeding years this animal would come to delight children of cross-plains immigrants, strike fear into the hearts of ranchers, and symbolize the conflict between utilitarians and conservationists. They would come to rival bison as icons of the Great Plains frontier.

. . . . .

The village of Lynchburg, Nebraska, nestles in a small valley several miles south of Clark's "Cupola." Outside Lynchburg, a metal marker inscribed with Clark's description of the first prairie dogs welcomes the back-road traveler. When I drove there in 2002, almost two centuries after Lewis and Clark had passed, local people called the pale pinnacle "The Towers" or "Old Baldy." From the gravel road I saw it as a small, chalk-colored bluff, remarkable only by contrast with the greener and gentler slopes surrounding it. Beyond and lower down, the Missouri curved lazily along the valley bottom.

The Lynchburg Ladies Sewing Circle, in anticipation of the forthcoming bicentennial of the famous journey, had taken to stitching prairie dog effigies from brown felt. According to the service station attendant who displayed them behind the glass counter that held the cash register, they sold better than patchwork quilts.

"Naw," he said, "there ain't no prairie dogs there no more. People talk of putting some back for the Bicentennial. But the land's private now."

He sold me one of the original "Lynch Dawgs." It still keeps watch with beady eyes from a high shelf above my books. I never heard whether Lynchburg boosters reintroduced prairie dogs to the vicinity of Old Baldy. Given the unrest generated among present-day agricultural interests and the state government of Nebraska by the mention of prairie dogs, I suspect not.

• • • • •

The discovery of prairie dogs heralded Lewis and Clark's entry into a different kind of country. The river and the route of travel had veered back west a few weeks earlier. The current quickened, and the voyagers climbed more rapidly in elevation. The air grew drier, the grass shorter. Two weeks earlier, Joseph Fields had killed the expedition's first bison, called buffalo by the chroniclers. As the party moved into present-day South Dakota, bison began to appear in herds of hundreds, then thousands.

On September 16 the party stopped on the west bank of the Missouri about ten miles north of where Interstate 90 now crosses the river in South Dakota. The future Fort Pierre and adjacent capital of South Dakota lay ahead of them sixty miles to the northwest. Here, well into bison country and what ecologists later would call the mixed-grass prairie, they spent two days and explored on foot the surrounding country. Their journals reek with the odor of the pristine. Animals they saw showed little fear of people. Said Lewis: "Capt. Clark and myself killed a buck each immediately on landing . . . the deer were very gentle and in great numbers . . . three hunters soon added eight deer and two Buffaloe . . . [which were] so pour that we took only the tongues and marrow bones. . . . Extensive planes had been lately birnt and the grass had sprung up and was about three inches high."

The next day Lewis strolled farther afield with six of his hunters. They encountered a flat plain about a mile wide and three miles long that was

> intirely occupied by the burrows of the *barking squril* heretofore described; this animal appears here in infinite numbers, and the shortness and virdue [verdure] of grass gave the plain throughout it's extent of beautifull bowlinggreen in fine order . . . [on which we saw] a great number of wolves of the small kind [coyotes], halks and some polecats . . . this senery already rich pleasing and beautiful, was

> still farther hightened by immence herds of Buffaloe deer Elk and Antelopes which we saw in every direction feeding on the hills and plains. I do not think I exaggerate when I estimate the number of Buffaloe which could be compreed [comprehended] at one view to amount to 3000.

Why the rising exuberance as these children of the eastern woodlands plunged deeper into the heart of the Great Plains? By this time, four months after their departure from St. Louis, the grueling pace of twenty miles and more in some days must have taxed their energies. Describing landscapes certainly had lost its novelty. Did their excitement reflect the shedding of agrarian shackles, escape from cultural containment, promise of personal gain, dreams of brown-skinned women? Or did they simply begin to sense the full extent of the new national domain, the far view of collective empire? Having likewise come from woods onto grass at a young age, I wonder whether their excitement was not a natural response of dispersing males encountering opportunity in the natural habitat of the species.

Lewis and Clark's journey spawned a succession of Anglo-American expeditions into and across the Great Plains during the next fifty years. Uniformly, the chroniclers of these expeditions had come from the forested parts of the country. Uniformly they marveled at the visceral appeal of the great grazing pastures, still free from the bawl of cattle and the yap of the herder's dog.

Major Stephen H. Long's party traveled west up the Canadian River in 1820. On August 28, about ten miles south of present-day Oklahoma City, party member Edwin James related how they climbed up out of the Canadian River valley to the open plain: "Here we passed a large and uncommonly beautiful village of the prairie marmots, covering an area of about a mile square. . . . The grass on this plain was fine, thick, and close fed. As we approached, it happened to be covered with a herd of some thousands of bison; on the left were a number of wild horses, and immediately before us twenty or thirty antelopes and about half as many deer. As it was near sunset the light fell obliquely upon the grass, giving an additional brilliancy to its dark verdure."

Twelve years later, the famous American novelist Washington Irving embarked from St. Louis with a group of adventurers. They traveled west and then south to the same region of the southern plains Edwin James had crossed. On October 21, 1832, about thirty miles north of present-day Oklahoma City, Irving could not control his eloquence: "After a toilsome march of some distance through a country cut up by ravines and brooks, and entangled by thickets, we emerged upon a grand prairie. . . . An immense extent of grassy, undulating . . . country with here and there a clump of trees . . . the landscape deriving sublimity from its vastness and simplicity. . . . The prairies of these great hunting regions differed in the character of their vegetation from those through which I had hitherto passed. Instead of a profusion of tall flowering plants and long flaunting grasses, they were covered with a shorter growth of herbage called buffalo grass."

At the approach of midcentury, on September 12, 1849, Captain R. B. Marcy saw on the headwaters of the Clear Fork of the Brazos west of San Angelo, Texas, "as beautiful a country . . . as I ever beheld. It was a perfectly level grassy glade . . . with . . . large mesquite trees at uniform distances. . . . The grass is of the short buffalo variety, and as uniform and even as a new mown meadow."

• • • • •

The Bad River Ranches—composed of several smaller ranchers bought and pieced together by Ted Turner—straddles the Bad River, South Dakota. Twenty miles downstream, the Bad River meets the Missouri near Pierre, the state capital. Lewis and Clark had a bad encounter with the Teton Sioux at the mouth of this river on September 24, 1804, and named it the Teton. Only later did cartographers change its name to the English word for what the Sioux thought about it. Indeed, it carries a lot of silt, looks muddy nearly all the time, and—although Clark in 1804 said it was seventy yards across two miles upstream—in droughts you now can walk across it in places twenty miles upstream without getting your feet wet.

Grass covers most of the flat to steeply rolling terrain of the Bad River Ranches, which lies in the heart of what ecologists call the northern mixed-grass prairie. The two dominant grasses on the ranch are the mid-height grass called western wheatgrass and the short, turflike buffalograss. To conventional livestock managers, a good stand of western wheatgrass signals good management; too much buffalograss means overgrazing. The ranch runs bison, not cattle.

On Bad River Ranches I got a chance to see first-hand the interactions of weather, grass, bison, and prairie dogs. When rain falls abundantly and bison graze lightly, western wheatgrass proliferates, suppresses buffalograss, and allows stalking predators to push back the margins of prairie dog colonies. Conversely, under heavy grazing and in drought years, western wheat suffers. Buffalograss and prairie dogs proliferate because the former gets a full dose of sunlight and the latter get a better view of approaching coyotes and badgers.

Mr. Turner wanted more prairie dogs at Bad River. During the first several years of the new millennium, biologist Kristy Bly lived on the ranch and fought tall grass and yellow clover to this end. Possessing extraordinary endurance, Kristy commonly worked from daylight to dark out on the prairie. She planted new dogtowns and watched them germinate. She measured colony growth and its response to bison grazing and precipitation. She brought hard numbers to the prairie dog-bison-rainfall equation. Under light grazing and moderate to high rainfall, prairie dog colonies shrank. With heavy grazing and drought, they expanded. Range conditions judged bad by conventional measure turned out to be good for prairie dogs.

Accounts written by nineteenth-century travelers across the Great Plains show that wild bison often "overgrazed" the range. Some travelers encountered vast areas grazed so heavily that little forage remained for their own horses and mules. The "pour" bison killed by the Lewis and Clark party on September 16, a week downstream from Bad River, might have reflected overstocking by Frederic Clements's standards. Under such conditions, prairie dogs prospered.

Prairie dogs usually graze even more intensively than bison. Over time in dogtowns, grazing-sensitive grasses decline and disappear, leaving only those tolerant of being trimmed very short. In the oldest prairie dog colonies on the Bad River Ranches, only buffalograss remains abundant. This species likely caused the "beautifull bowlinggreen" appearance of the colony Meriwether Lewis wrote about some sixty miles to the southeast of Bad River on September 17, 1804. Buffalograss also probably dominated the shortgrass scenes described farther south by Edwin James, Washington Irving, and Captain Marcy.

. . . . .

My James Beard turf manual describes forty species of turfgrasses used in the United States. Most are cool-season species from Eurasia, and these cover the majority of the lawn and sports-turf acreage in America. Bermuda grass and a few other warm-season species, likewise alien, prevail locally in the South and Southwest, usually in combination with one or more cool-season varieties to keep the turf green in winter. Only six species Beard mentions came originally from the United States.

The four most important of these six are Great Plains grasses: the cool-season western wheatgrass and the warm-season blue grama, side-oats grama, and buffalograss. James Beard, turfgrowers nationwide, and university agronomists agree that buffalograss tops the acreage of native turf hands down. One might guess that American patriots would prefer buffalograss over alien species, but I've found no evidence to support this.

Buffalograss shows one attractive quality most imported species lack: drought hardiness. I see it flanking some city streets in the arid Southwest and growing naturally in some close-mowed parks beside highways in the western Great Plains. University extension specialists from Texas to the Dakotas recommend groundskeepers use it if they want to minimize water and fertilizer use on golf fairways, parks, and lawns. In areas with looming water shortages, some city and county ordinances favor its use as well.

One Saturday morning in early June 2006, I helped two of my neighbors plug buffalograss. Bob Robertson and Mary Newkirk didn't wear American flag pins, but they showed other good American tendencies. They'd grown up in metropolitan California and did well at free-market endeavors. Several years ago they bought a tract of raw land near my town and, with minimal help from contractors, designed and built their house. Mary planned the grounds, and she wanted a croquet lawn of buffalograss. (I don't know who would have been more offended, the French or the buffalo.)

Plugging grass means setting out rooted wedges. When I arrived Bob and Mary had prepared the ground according to instructions from the grower. The soil stood weedless, loamy, and wet. They showed me the plastic flats delivered by UPS from a Nebraska grower. Each flat contained a couple dozen close-spaced pockets from which thrust the curly heads of buffalograss clumps, often bound to each other by stolons, or runners.

Bob used an electric drill with an inch-and-a-quarter spade bit to punch holes for the plugs. With California-vineyard regularity he lined the holes up arrow-straight east, west, and diagonally. I separated the wedges by cutting the intertwining stolons with yard scissors, and sliced lengthwise the lower part of each root mat to redirect it downward. Mary plopped the wedges into the holes.

By noon we could see why experts recommended seeding rather than plugging for large plantings. The ground we'd covered looked pitifully small. Bob's lines began to meander. Mary brought iced tea. I thought about planting a trial patch in my own yard—a smaller plot, with plugs farther apart—to see if buffalograss could compete with the African bermuda that had crept in through benign neglect.

As I drove away Mary and Bob went back to work with that determination that had brought them from the fleshpots of California to the last frontier of the American dream. Across the fence, on U.S. Forest Service land, a pasture of blue grama advertised its even greater resistance to drought but unfortunate tendency to grow in clumps. It would be hard to play croquet on blue grama.

TEN Taming of the West

Figure 10. Emigrants Crossing the Plains by Felix Darley. Courtesy Yale Collection of Western Americana, Beinecke Rare Book and Manuscript Library.

Few things stir the imagination of the human male as does the mythical odyssey. Young man leaves home, travels far, endures dangers and temptations, and finally returns home triumphant. The legendary Greek hero Odysseus gave name and plot to the first major epic of this genre. When young he left home to endure the siege of Troy, then started back across the Aegean Sea only to blow off course in a storm. After years of adventure in strange lands he returned safely home to the arms of his wife, Penelope.

Young men probably have been moved by legends of travel and conquest since the birth of language. In the centuries following Odysseus, countless stories featured the same plot, with different characters. In all corners of the world young people—largely men until recently—have searched beyond home for better places, more desirable mates. Story, song, book, and now video call to a never-ending stream of youth to venture over the next mountain, to the other side of the homeland sea.

The odyssean epic makes legend of the evolutionary phenomenon of dispersal. The dispersing male leaves home and community in search of vacant habitats and suitable mates. He encounters new places, dangerous predators, male competitors, female temptresses. If he survives, he finds a place to settle down, triumphant to the extent that he begets young and inspires his own male progeny to embark upon the age-old journey.

The American odyssey found its most glorious stage on the grasslands of the West. Here, to greet adventuring Euro-Americans in the nineteenth century, lay the majesty of the long view, the thundering herds, the sanctuary beside stream and bluff. Behind the westering traveler lay the thick woodlands and crowded farmlands of Europe and America. Beyond the forward horizon beckoned the ancestral human habitat, a landscape that the American hustler of Manifest Destiny advertised as essentially vacant, waiting for the rightful heir.

For more than a century now, horsemen and horsewomen of the American West have stirred the imaginations of millions, not only in America but around the world. People visit dude ranches, attend rodeos, read western novels, or watch movies in vicarious indulgence. The image of the western pastoralist has proved durable far beyond economic utility and common sense. Rodeo riders continue to compel admiration despite a hundred-year history of failed livelihoods and crippled bodies.

• • • • •

In Larry McMurtry's fictional odyssey *Lonesome Dove,* dispersal comes late in life for Gus McCrae. He and his partner, Woodrow Call, having endured the Civil War as Texas Rangers, succumb to the siren summons of a thousand-mile cattle drive from Texas north to Montana. Call anticipates limitless grazing pastures on the northern range; Gus has his mind on women and other more immediate rewards. During the long journey north they survive the hazards of nature, evil men, and Indian raids. The aging Gus renews acquaintance with an old flame, but then near the end of the trip gets shot by Indians. Call eventually finds the mythical habitat and starts a younger man on his own quest. Finally they return home, the faithful Call packing Gus in a plaster coffin.

In real life, Charles Goodnight and Oliver Loving pursued a McMurtry-like journey that epitomized the domestication of America grasslands. Goodnight endured the Civil War as a Texas Ranger fighting mostly Indians, the older Loving as a cattle drover supplying the Confederate Army with beef. In 1866, two years after the war ended, they joined forces in an economic venture to drive a herd of cattle from their home country near present-day Weatherford, Texas, to New Mexico. They had heard the U.S. Army and Indian agencies at Fort Sumner paid well for beef.

To get to Fort Sumner, they took their cattle along the ten-year-old Butterfield Overland Stage route west from Weatherford. They struck the Pecos River in far West Texas at a place called Horsehead Crossing, south of present-day Monahans. Here they turned the cattle north, upriver, on into New Mexico. Eventually they reached Fort Sumner, out on the plains a hundred miles east of Albuquerque. Goodnight turned thirty that year, Loving fifty-four.

Their first drive proved so lucrative they conducted a second, then a third, all in less than two years. On the last trip Loving went ahead into New Mexico and got into trouble with a band of Comanches. Wounded under a streambank, he sneaked out at night and caught a passing wagon to Fort Sumner. There he died from infection, and they laid him under New Mexico dirt.

Goodnight later had the body exhumed and carted back to Weatherford. Historians say Loving traveled home in a bed of charcoal covered by a tin shell made from emptied cans. The Loving name still looms large on the geography that claimed his life.

But Charles Goodnight's journey had only just begun. After Loving died, he drifted, Odysseus-like, under a fickle wind. For the next few years he drove cattle to Denver and the Colorado gold fields, farther north to Cheyenne, and to the railhead at Dodge City, Kansas. Then in 1870 he returned home and married his own fair Penelope—Molly Dyer, lifelong sweetheart. But he could not stay.

Having seen wilder lands and better prospects than Weatherford, Texas, he soon moved wife and belongings to Pueblo, Colorado. Here he set up a ranch at the western edge of the Great Plains. He ran cattle on arid shortgrass for six years, went broke, and began to look afield for greener pastures. The timing and direction of his next move gave him a key role in taming the last wild grazing pastures of America.

In the spring of 1876, he drifted south from Pueblo and struck the Canadian River in northeastern New Mexico. Before him dusted a herd of Colorado cattle. Turning east down the Canadian, cattle and men wound up summering in the Texas Panhandle. That fall, under the eastern rim of the High Plains cap rock, they found the perfect spot, the journey's end.

Spanish-speaking adventurers ranging east into Texas from New Mexico already knew about the place. They called it Palo Duro Canyon—valley of hard sticks. In these grassy breaks southeast of present-day Amarillo, abundant springs from the cap rock above watered the cottonwoods and bison below. Here in the canyon, on the Red River headwater stream called Prairie Dog Fork, Goodnight selected a spot to build what would become his famous Home Ranch.

In the vast surrounding prairie, Hispanic sheepherders, bands of outlaws, and Comancheros—New Mexicans who traded with the Comanches—coexisted with the Indians. The Home Ranch proved the first of many Anglo ranches to come in quick succession as American frontiersmen brought to climax the play of Eminent Domain and ousted the last of those who came before.

On first bringing his cattle off the cap rock to the Home Ranch, Goodnight came upon a herd of about ten thousand bison. As his crew spread the cattle out over the surrounding range, they chased away thousands more bison to "save the grass" for the cattle. Often the cattle mingled with the bison.

On a still day Goodnight's cowboys sometimes could hear the boom of buffalo rifles in the distance to the east. The hide hunters had found their own paradise.

• • • • •

Josiah Wright Mooar drove his wagon down to the Texas Panhandle in the fall of 1873, three years before Goodnight came. On the seat beside him sat not a wife but a ponderous rifle. Wright, as his friends called him, sought bison hides for market. He had started hunting bison three years before to supply meat to Fort Hays, Kansas.

Indeed, the business of hunting bison for their hides had started in Kansas about 1870. It filled the gap left when the supply of cowhides from the Argentine pampas to America suddenly dropped off in the late 1860s. In that pre-oil era, leather had critical functions—footwear, harnesses, saddles, machine belts—and few good substitutes.

Nineteen at the time, Wright Mooar entered the bison hunters' ranks at a malleable age. Rural fathers and military recruiters alike know it's easy to teach most young men to kill just about anything that moves. So it was with Wright. He took an old Civil War Springfield rifle out on the prairie and started killing bison.

The railroad, necessary to ship hides to eastern markets, had reached Fort Hays a few years earlier. After Mooar and other shooters laid waste to the Kansas bison, skinners ripped off the hides and the laden wagons rumbled away to the railhead at Fort Hays. By spring of 1873 bison numbers in Kansas had shriveled to a pitiful remnant of their former glory. The hunters looked around for another supply.

Texas held the only large herds of bison left in the southern plains. Hunters hesitated to go there because they heard rumors that federal agreements with Comanches and other tribes prohibited it. They thought Army patrols would seize any wagons moving south into the 34-mile-wide "neutral strip" between Kansas and Texas.

A group of hunters near Dodge City, seeing the decline of Kansas bison and the approach of the railhead to Dodge as a possible shipping point for Texas hides, decided to settle the rumors. They sent Wright Mooar to meet with the head man at Fort Dodge. The man proved to be Major Dodge himself. Natural historians pondering the wherefore and why of bison decline later would pore over the major's chronicles.

Right away, Mooar made a hit with Major Dodge. While the seasoned young hunter rambled on about buffalo and buffalo hunting, Dodge took notes. Then Mooar posed the question: What would the Major do if a bunch of hunters went down to Texas? The reply proved eminently satisfactory to Wright and the waiting hunters: "If I were hunting buffalo, I would go where the buffalo are."

According to Mooar himself, he made the first wagon track to the new hunting grounds. Close on his heels a flood of hunters soon poured down from Fort Dodge to Texas.

The gun Wright Mooar kept close at hand as his wagon bounced south had a long barrel and a short history. He claimed to have had a hand in its development. After shooting Kansas bison with the old army Springfield, he sent a letter to the Sharps rifle company back in Connecticut saying that buffalo hunters needed a longer-range and more powerful gun.

By 1873, the rifle company's response lay in the hands of Mooar and a growing number of other hide hunters. The Sharps buffalo rifle, a monster that weighed twelve to sixteen pounds and hurled .44 or .50 caliber bullets to kill bison and Indians at a distance of up to a half mile and more, would make history. In the six-year wink of time between 1873 and 1879, Wright's "Big 50" and others like it would permanently cripple the southern plains Indians and take out the Texas bison.

The Sharps began to seriously weaken the resolve of the Comanches the summer after Mooar came down from Kansas. On June 27, 1874, about five hundred Comanche, Kiowa, and Cheyenne warriors under the famous Quanah Parker attacked a company of twenty-seven hunters and a supplier and his wife at Adobe Walls. The hide-hunting party quickly took shelter in sod huts—made from buffalograss squares peeled off the prairie—at this primitive outpost about sixty miles northeast of Amarillo. The Indians carried short-range Winchesters at best. The hunters repelled the initial charge and began picking off horses and Indians at frightening distances.

Day by day, more hide hunters sneaked in from the surrounding prairie. Bands of Indians moved into and out of sight, farther away now after having seen the capabilities of the buffalo guns. Several days after the initial attack, a hunter named Billy Dixon turned the tide permanently: he dropped an Indian at a distance he later claimed to have measured at over 1,500 yards, phenomenal even by today's standards of rifles and marksmanship. The warriors drifted away. Superior technology had given a sampling of its power to come.

Mooar and other hide hunters normally shot bison from a "stand," or vantage point, three to four hundred yards away. They learned to place their shots behind the rib cage and aim first for the "lead cow" of a herd so as to not spook the remainder. The rifle would boom and the confused bison would drop, one after another. Afterward the blood-caked, grease-covered skinners would come with their mules to rip off hides and load them onto wagons.

The carcasses, occasionally minus a tongue or steak to feed shooters and skinners, would swell in the sun. The turkey vultures would circle lazily and land. The wolves, smelling the bonanza from afar, would assemble to feast. On warm days vast swarms of blowflies would spiral up from rotting flesh.

Methodically the shooters conducted their work. More like terminators at slaughterhouses than hunters, some kept track of the number of bison killed. A notch per animal soon would have whittled away entire gunstocks. Among the more persistent shooters, record kills exceeded 300 a day and 5,000 for a winter season. Mooar himself claimed to have killed 20,500 bison during his ten-year career. Once, he said, he shot ninety-six from a single stand, eight or nine miles north of where he finally settled down in Snyder, Texas.

Within a few years after the first hunters and skinners came down from Fort Dodge, suppliers from Fort Worth likewise began to gravitate toward the Texas hunting grounds. Loading their wagons with flour, lard, powder, and bullets, they guided their mules west and set up shop at Fort Griffin, near present-day Abilene. By the winter of 1876–77, the westering railroad had reached Fort Worth and the number of hunters west of Fort Griffin had swelled to 1,600 or so. Such a gathering promised security from Indians and assured the doom of the bison.

Look at a Texas map. The herds on which the hunters converged fed mostly in the region west of Abilene and Fort Griffin, north of Sweetwater, east of Lubbock, and south of Clarendon. They ranged as far west as Goodnight's Home Ranch and the cap rock that loomed above it. Here, a quarter century earlier, on the headwaters of the Clear Fork of the Brazos, Captain Marcy had beheld the wonder of the shortgrass savanna, one of the last wild grazing swards in America, and remarked on the unexcelled beauty of these buffalograss pastures. Over this same terrain, in the scant five years between 1874 and 1879, hide hunters in pursuit of fleeting fortune would obliterate the capstone of a hundred thousand years of grazer evolution.

As the spring of 1879 lengthened, the few hunters that were left poked around for clots of survivors. They found the range oddly vacant of the sounds and smells so recently rampant. Like so many before them, they had strangled the object of their passion. Already, cattle belonging to Goodnight and others bawled from the greener patches near water. The quickness and starkness of change had been little short of miraculous.

The hunters scattered. Many no doubt made for Dodge City or Fort Worth. Here perhaps they washed the accumulation of prairie dust and bison fat from their bodies and spent their earnings on barroom games and backstreet temptresses. Some drifted on to other killing fields—haunts of grizzlies, wolves, and even prairie dogs.

A few hurried north to participate in the slaughter of the last large bison aggregation in existence anywhere, in Montana and the Dakotas. As the railhead reached Miles City, the hunters fanned out over the surrounding prairie. In a now familiar ritual, rifles boomed, bison fell, and hides headed for train yards. After Texas, the smaller harvest of the northern herd seemed almost anticlimactic. By 1883, this herd had followed the southern one into oblivion.

Why did Wright Mooar settle down on the killing fields? Did he live out his days telling stories to big-eyed children on the town square of Snyder—stories of how the bison had blackened the prairie, of how many he had shot, how much money he had made? Why he stayed we may never know, but people said that, before the last shot echoed across the prairie, he claimed the Holy Grail of hide hunters: he killed a rare white buffalo. The stars had come into alignment. Another Odysseus had come home.

How big was the slaughter overall? Historian Andrew Isenberg estimated the southern herd may have been upward of 15 million strong when young Wright Mooar came out onto the bison plains from Vermont in the late 1860s. Major Dodge later estimated, on the basis of hides shipped from railheads, that hunters killed more than a million southern bison a year during the peak years.

Bison shot but not skinned, hides rotting and not hauled in, and inferior hides left at shipping points added to the numbers. Cattle diseases—Texas tick fever, bovine tuberculosis, and others—probably took a substantial toll as well, unaccustomed as the bison were to these European scourges. Disease probably hit hardest in Texas, where, in contrast to most other parts of bison range, cattle of the first ranchers mingled extensively with the bison. Regardless of immediate cause, the final demise of the American bison can be laid directly at the door of Europeans intent on replacing the totem of the native peoples with that of their own.

• • • • •

The Texas harvest of hides peaked just as Goodnight settled onto his Home Ranch, and it ended two years later. Once the hunters had packed up their greasy bedrolls, their skinning knives, their Sharps rifles, and moved on, wagons from Fort Worth plied the plains for bones. Bison skulls, leg bones, ribs, and vertebrae grew into small mountains at the Forth Worth shipping yards, then left by the trainload for fertilizer factories and china factories around the world. My wife owns an old set of bone china inherited from her grandparents in Michigan. Might it contain molecules of Texas bison?

By 1878, the transition to domestication on Goodnight's ranch seemed well on the way to completion. But wait. His wife Molly, distraught at the carnage strewn about by the hide hunters, came to Charles with a plea: "Save some buffalo. Bring me some babies."

And so he did. He searched about with some of his ranch hands and found orphaned calves here and there. They roped ones small enough to handle, hauled them in to ranch headquarters, and suckled them on cattle. Thus began the famous Goodnight buffalo herd, which pleased Molly and helped save from extinction the icon that in time would decorate flags, coins, and dioramas across the country.

After the bison slaughter, Goodnight teamed up with a rich Irishman named John Adair to buy more land in and near Palo Duro Canyon. Like many later successes on the western range, he tapped into foreign capital and bought the best sites first—those with good water, good grazing, trees, and good building places. This high-grading of the range gave him control over vast additional acres. Goodnight survives in history and legend as the quintessential American Rancher, the kind of man some big-sky mothers still want their babies to grow up to be.

• • • • •

When scientists in my profession publish results of their investigations, they try not to reference other studies more than twenty or thirty years old. Those works are considered outdated, superseded by later investigations presumed more worthy because more recent. Like the echoing mantra of Eminent Domain, the science that guides us seems anxious to discard the old, bring in the new. This casting aside of history may partly explain why, the older I get, the more I'm fascinated by it.

An old book I have describes the invasion of the Great Plains by cattle. The book smells like the libraries I inhabited during my dispersal years and where I first discovered girls who liked reading. Does that bias my appreciation for it? Published in 1950 and authored by Marion Clawson, then director of the U.S. Bureau of Land Management, it carries the title *The Western Range Livestock Industry.*

Four of Clawson's maps show the western United States decorated with dots. The maps snapshot the years 1860, 1880, 1900, and 1920, with each dot representing 2,000 cattle. The moving of cattle into the Great Plains between 1860 and 1920 looks on the maps like colonies of bees swarming north from Texas.

In 1860, the year the Civil War began, the eastern third of Texas claimed nearly all the cattle in America's interior grasslands. Historian Walter Prescott Webb placed the number of Texas cattle at the time at about five million. By stark contrast, the remainder of the Great Plains at this time had only a minor sprinkling of dots, mostly close against the Mississippi River in the Midwest and with a few lonely outliers in Montana and Wyoming. Texas cattle numbers escalated during the war while the menfolk were off shooting at Yankees and Indians.

By 1880, satellite swarms had moved north and west into the country Goodnight had settled and the hide hunters had cleared of bison just a few years earlier. Smaller swarms had lodged along the eastern edge of the Rockies from northern New Mexico to southern Wyoming—a meat source for the mining industries clustered there. The great cattle drives north from Texas had as yet made little mark on the vast rangelands of the central and northern plains.

The cattle drives matured into an American epic in the early 1880s. Texas, not the eastern states, continued as the primary source of livestock to Kansas, Nebraska, and the Dakotas. The vast northern plains sucked in cattle as a sponge absorbs water.

In the mixed-grass prairie of the central and northern plains, an unprecedented range vacancy prevailed for a few decades between the time bison disappeared and equivalent numbers of cattle replaced them. Observers at the time remarked on the recovery of the grass. Professor B. F. Mudge, first state geologist of Kansas, for example, noticed that "the buffalo grass was naturally displaced by taller nutritious grasses soon after the buffalo disappeared." Naturalist W. A. Bell traveled across Kansas in 1867 and also noted cause and effect: "Doubtless no grass [except buffalograss] could bear so well the heavy tramp of thousands of buffalo continually passing over it but . . . as settlers advance, and domestic herds take the place of big game, the coarser, more vigorous, and deeper-rooted grasses destroy it, and take its place."

Before the "domestic herds" multiplied and reversed the trend, many other observers likewise went out on the prairie and saw what was possible with little or no grazing. Like John Weaver later would do in Nebraska, they saw grass through the eyes of the haymaker and his cow. It looked better now that the bison were gone. These utilitarian men needed little nudging to embrace climax as a phenomenon waiting for a name and benediction.

By 1900 the cattle drives had essentially ended. The blackest swarms of dots on Clawson's map had shifted from Texas to the Midwest—to Iowa and eastern Kansas and Nebraska. Most of the prairie in these places, largely tallgrass, had been plowed under for corn. Cattle thrived on corn. Farther west in mixed-grass and shortgrass country, the Charles Goodnights of the range, subsidized by eastern and European money, had pretty much claimed and stocked up their cattle empires.

In the wink of an evolutionary eye, man and dominant beast crossed the cultural chasm from wild to tame in the American heartland. Along the way, out of hubris and habit, the domesticated killed, pushed out, or converted the undomesticated. By such extermination, they burned the bridges of their crossing. Few among the triumphant mourned. After all, why would anyone want to travel backwards?

ELEVEN Production Science Comes to the Range

Figure 11. Cattle production, Nebraska Sandhills. Photo by author.

The dust of the last cattle drive from Texas to Montana had scarcely settled and the last plains Indian sent to reservation when the trickle of westering Anglos turned into a flood. Free land drew Civil War survivors west like blowflies to a skinned bison. Surging onto the central grasslands came people dedicated to food production, European style. The farmers among them swarmed out over the tallgrass prairie and moister parts of the mixed-grass and, by the end of the nineteenth century, occupied nearly all the regions wet enough to grow the corn or wheat varieties of the time. The drier regions farther west filled with ranchers, cowboys, cows, sheep, and horses.

The throng of prairie invaders resembled miners more than agriculturists. Following the lead of the hide hunters before them, they seemed intent on getting theirs while the getting was good. Perhaps disillusioned by war, most seemed to hold remarkably short visions of the future. Men with money and men with none fought among themselves to lay claim to the best chunks of land. The word *conservation* had not yet made its way into their language.

Railroads branched out to gather the new products of prairie sunshine for market. The wetter soils began to bleed their fertility away in loads of corn and wheat bound for Chicago and the world. Range cattle likewise fed mostly the people in faraway cities. Trains heading east bulged like pulsing arteries.

A century later, in a short 2001 *Rangelands* article titled "From the Dust of Shame: A History on How the Profession of Range Management Was Born," eminent range scientist Thad Box told the story of that initial exploitation of the range and the slow emergence of sensible management. He described the bloating of the Great Plains with cattle in the early 1880s and the stupendous "die-ups" from blizzard and drought later that decade and into the next. Depending on location, 30 to 90 percent of the range cattle died in a ten-year period, and for brief stretches in 1884, 1886, 1887, and 1894, the rangeland scavengers lived very well.

The U.S. government responded quickly, unlikely as that now may seem. In 1895, the Department of Agriculture set up two federally funded range experiment stations, one each in the Texas plains towns of Abilene and Channing. Federal agents found a handful of ranchers to collaborate in an experiment that showed the consequences of too many cattle. They convened a meeting of stockmen in 1897 to spread the gospel, but encountered ridicule. According to Thad Box, one rancher offered a resolution: "None of us know, or care to know, anything about grasses. . . . We are after getting the most of them while they last."

The basic ideas for applying science to range management go back as far as 1870 and the American Midwest. That year an ambitious young instructor in botany named Charles Bessey arrived at the newly opened Iowa State Agricultural College. The same year, not far away in Kansas, Wright Mooar started his career as bison hunter. The stage was set to take out the old, usher in the new.

Bessey moved in 1884 from Iowa to the University of Nebraska. There he taught such American notables as novelist Willa Cather and jurist Roscoe Pound. But his influence on the world of grass would come largely through his student and disciple Frederic Clements. The same year Thad Box's rancher proclaimed his "eat out and get out" philosophy, Clements graduated and began teaching his own philosophy of grassland conservation at the University of Nebraska.

• • • • •

Late one semester at Texas A&M in the early 1960s, a professor of range management introduced me to the value of uniformity on rangelands. By that time I'd accepted uniformity as a way of life on lawns, cornfields, and pine plantations. My father's instructions for managing family food production had smacked of it: rows three feet apart and plumb-bob straight; black-eyed peas planted two feet apart and two inches deep; one tablespoon of 5-10-5 per hill; no weeds, no grass.

The professor instructed us to prepare a range management plan. He made it clear we would need to pipe water, fence range, and do something about brush encroachment. I hesitated, uneasy with the utilitarian cadence of his chant. He seemed to be asking us to convert Zane Grey country into a kind of farm.

"This will count heavily in your class grade," the professor said. "Look, you'll take this hypothetical ranch in north-central Texas, managed poorly to date. Hilly, water mostly ephemeral and only in ravines, lots of mesquite, and mixed breeds of cattle dropping calves year round. You get the picture. Here's a map and description. Make it productive, economically sustainable. Due in a week."

Having spent almost three months in his class, I knew what he wanted. So in my plan I took out a loan from the bank in Hardscrabble, Texas. I piped water from the few good springs and wells on the ranch to steel tanks nearby, and built dirt tanks elsewhere with development aid from the federal government. Soon I had permanent stock watering stations no more than a mile apart. I distributed salt blocks on ridges to pull the cattle away from water to places they wouldn't normally graze. I built cross-fences to influence when and where cattle would graze. I teamed up with a charcoal-maker to cut and haul away mesquite, the principal competitor with grass.

On paper I soon had pure-bred Herefords grazing uniformly from fence to fence, calving in unison in February, and feeding dependably on government supplement in drought years and cold winters. Knee-high sideoats grama, Texas wintergrass, and little bluestem rippled across the flats. I kept the grass near climax by reducing the number of cattle or providing supplement as needed.

The range professor liked my plan. My wildlife professors would not have liked it. They favored white-tailed deer, wild turkeys, and quail, animals that need habitat diversity. "Intermingle the habitat types," they would have said. "Keep some brush and trees for cover, roosts, and different kinds of food. Create edge. Aim for a savanna landscape."

My range management plan built on the notion of community succession and climax as laid out by Frederic Clements. I knew climax as the stable state a community eventually reaches absent disturbance. The Nebraska prairie study areas where Clements developed his notions of succession and climax had few or no grazers. Thus the grass stands produced maximum yields of hay and looked very good to cattle growers.

It turned out, coincidentally or not, that Clements judged the ecological condition of the range on a scale of supposedly absolute value that in fact more or less reflected the value of the range for raising cattle. Naturally the scheme appealed to American ranchers and indeed stock-growers worldwide. For most of the twentieth century, Clements's succession theory rode high on the range as the sole determinant of good and evil.

But back during the first decades of the twentieth century, few out on the range had yet heard of Clements, and cattle kept the range condition pretty low by his standards. The federal government set out to educate folks. They established more range experiment stations to study grass and sustainable livestock production. In 1934, largely on the basis of work by Clements, Weaver, their colleagues, and the experiment stations, Congress passed the Taylor Grazing Act. Its stated objective was "to promote the highest use of the public lands . . . [that are] chiefly valuable for grazing and raising forage crops."

Production science had come to the range, at least in concept.

• • • • •

It is now summer of 2007. From my back yard at about 5,000 feet elevation I get a good view of the greatest challenge to "sustainable livestock production" encountered by range scientists in the twentieth century: brush.

Near at hand on the hills beyond the creek, juniper and scrub oak intermingle with grass, thinly on south-facing slopes and more thickly on north-facing ones. Higher up this west end of the Mogollon Mountains, pinyon pine and mountain mahogany join in, fighting for the sun. You sometimes have to crawl to penetrate the thickets. Brush has moved in.

How do I know brush has encroached? I know by word of my neighbor, who ranched a lifetime on the mountain. "Used to be good grass," he said before he died several years ago, "all the way to the top. We drove stock—sometimes sheep in those days—up to higher country on trails now choked out by brush. In a lot of places now a cow has to eat bushes or die."

I know also by the thickets of young junipers clogging the spaces between stumps of the old ones cut ninety years ago for stoking the mine fires of the town of Mogollon. Juniper wood resists rot, and the massive axe-hewn remnants lie far apart, advertising a once-open savanna that's now a closed forest. Hike among the foothills and soon you see how the grass cover shrinks as the juniper cover expands.

Scientists apply measurement to test their convictions, and in the summer of 1993 I went out with a young scientist set on quantifying brush encroachment. Mark Miller had conceived a research project set in Catron County, New Mexico, as the focus for his master of science degree at New Mexico State University. He would measure the vegetation to test what both dogma and casual observation suggested: that brush had thickened greatly in the past hundred years.

We camped on Negrito Creek east of the village of Reserve. Pinyon and juniper trees cloaked most of the terrain; grass flourished on some flats and south-facing slopes. Negrito and its main tributaries rushed down the bottoms of steep and rugged canyons. Ponderosa pines reached high from canyon bottoms and steeper north slopes. The size ranges of trees of all species indicated few oldsters, many youngsters.

In his hand Mark carried a piece of new technology he called "the gadget." Even city navigators use them now. Called Global Positioning Systems, or simply GPSs, they receive signals from earth-circling satellites and translate the signals into coordinates on the earth's surface—your location. By today's standards, Mark's gadget was quite primitive. But it was the best tool available at the time.

Mark wanted to locate old land survey markers, which had been laboriously planted at section (square-mile) and quarter-section corners a hundred years or so before. Now they were lost, out in the brush and grass, but, Mark said, we should be able to find them with the gadget. The excitement of hunting Easter eggs paled by comparison.

A persistent problem came to light as we searched for the first marker. The GPS told us we should be within fifty yards of the metal stake with brass cap. We scanned about through the near screen of juniper and scrub oak. No marker in sight. We then ranged out in ever-widening circles. Thirty minutes or so later we found the marker, fifty yards away. I never fully understood what caused the frequent discrepancy between GPS readings and marker locations.

Other problems arose. Markers uncannily seemed to prefer rugged terrain. Was the one we sought at the base of this two-hundred-foot cliff, or on top? Did the original surveyor offset the marker to avoid that rock outcrop? Was the satellite signal bouncing off canyon walls and fooling the gadget? How in hell did the early surveyors come anywhere near being accurate in this kind of country?

One piece of information Mark sought at survey markers came in the form of a tree. The original surveyors had recorded in their notes—which Mark had copied—the distance and direction to the nearest large tree. Had new trees grown up nearer the marker in the last hundred years? Had the original tree died?

On occasion we could make out numbers the surveyors had etched on these "witness" trees. The glyphs, affixed by metal stamp on chest-high trunks skinned of bark, persisted in readable form only on alligator junipers. Though perhaps less evocative than petroglyphs, these stamps impressed us as testimonials to history.

Mark used additional measures to round out his research. Estimating the densities of trees of different ages and species, he looked for patterns in what he called "age structure" of the forest. He also located a series of mid-1930s aerial photos of the Negrito Creek watershed in the U.S. Forest Service office at Reserve and compared tree canopy coverage on them with coverage on similar photos taken sixty years later. He pored over reports written by early-day foresters and latter-day ecosystems analysts.

When he had finally looked at all the photos, "crunched" all the numbers, and steeped himself in the knowledge accumulated by others, he had an answer. Woody plants indeed had increased in number and cover over the last hundred years, drastically in some sites, less so in others, but inexorably. Few old-timers who knew about his work showed much surprise. Yes, they had been right all along.

But then Mark ventured onto more contentious ground. Why had the trees and brush encroached? Why had thickets of pine, juniper, and oak appeared suddenly during the past century but not before? As usual on variable landscapes, the potential reasons may have varied from place to place, said Mark the careful scientist. But the evil perpetrators seemed hard to mistake once he had stripped away the mask of "confounding factors."

One villain turned out to be grazing. It worked this way: Before sheep and cattle had come to the Negrito, the grasses had grown tall and lush. No grazers large enough to keep the grass trimmed roamed the landscape. Every five to ten years, on average, the tall grass—called "fine fuels" by foresters—had carried wildfire, the nemesis of the woody plant and the friend of grass.

For centuries fire had kept one-seed and Utah junipers, pinyon pines, and oak brush confined to ravines, saddles, north slopes, and other locations not vulnerable to fire. Elsewhere, ponderosa pines and alligator junipers had spread themselves thinly, thus preventing crown fires with sufficient heat to kill the occasional tree that made it to maturity. The grass fires burned cool, decade after decade, weeding out the small woody sprouts, leaving some of the larger, and nurturing the grass. A wide-open savanna prevailed.

Then in the 1880s came European grazers—sheep and cattle. Year after year they consumed with great efficiency the fine fuels that once had carried fire. Freed from fire, the shrub and tree seedlings survived and proliferated. The forest began to close. The grass, beaten down by unaccustomed grazing and shaded out by encroaching brush, retreated.

But there's another likely suspect, cautioned Mark: climate change. Several lines of evidence suggest that the prevalence of wetter-than-usual winters and droughty summers in the twentieth century may have favored the survival of trees over grass. Juniper, pinyon, and ponderosa pine suck up winter moisture to their advantage. Dry summers whittle away the advantages of grass and diminish the fine fuels that carry fires.

Woody plant invasion likely resulted from a combination of these forces, Mark concluded. Grazing and climate change colluded to set in motion a monumental change that continues to diminish the carrying capacity of the range for cattle. And there's no sign that the trend will reverse itself.

• • • • •

Ever since the coming of Europeans, some version of Mark's story has repeated itself in just about every corner of the pastoral West. The recent book *Brush Management,* edited by Wayne T. Hamilton and colleagues, ranks woody plant control as one of the most expensive endeavors of the ranching business. The troublesome species vary with location: creosotebush in the arid Southwest, pinyon and juniper at higher elevations across much of the West, mesquite in the southern Great Plains and Southwest, sagebrush in the Great Basin and northern Great Plains, eastern red-cedar in the southern tallgrass and mixed-grass prairies, and so on.

Reasons for brush encroachment vary, but, as Mark concluded, cessation of fire tops most lists. On many of the more arid ranges, grazing and the fire reduction that results from grazing stand as the prime collaborators in brush encroachment. In higher-rainfall prairies, where more people live and grass sufficient to carry fires sometimes persists despite grazing, fear of fire borders on the paranoid and firefighters rise to hero status. Regardless of location, range fires generally don't mix well with people anymore and can't be called upon to fix the brush problem.

What to do? In my range plan I offered a solution for mesquite encroachment, but in real life the problem proves less tractable than I imagined. The weeding of woody plants from the range—commonly called brush management—usually employs mechanical, chemical, or biological methods. Controlled burning works in some areas. Application of science has refined each over the years.

Mechanical brush control started in America with cutting or grubbing by hand, hardly a scientific endeavor. As a kid I grubbed yaupon stumps from our front ten acres, a technique already more than a hundred years old. Laborers on the Jornada Experimental Range in southern New Mexico hand-grubbed over four thousand acres of mesquite as late as 1958. In the summer of 1962 I worked in central Texas where government aid subsidized a marginalized work force known locally as "cedar-choppers" to remove thickets of Ashe juniper with double-bladed axes. The cedar-choppers seemed to have little schooling and no permanent homes, but moved their families from job to job and lived in tents.

Bulldozers and tractors, having first crawled onto the range in the 1930s, soon sent manual control into retirement. They gave rise to a combination of methods applied from the seat of the machine: blading, shredding, cabling, chaining, roller-chopping, disking, plowing, and railing. Today bulldozers still dominate the field in ever more glamorous renditions: wheel-loader grubber, tractor-loader grubber, excavator grubber, skid-steer loader, regrowth plow, and self-propelled shredder. I sometimes wonder what happened to the cedar-chopper kids whose parents got displaced by the machines.

World War II brought the chemical age to agriculture. On the range as on the farm and in the drugstore, postwar chemistry is where science really found its niche and corporations really got rich. The same folks who gave us DDT during the war gave us shortly thereafter 2,4-D, 2,4,5-T, sodium arsenate, and a host of other novel poisons derived mostly from oil. At the time, chemicals seemed the way to go—effective, cheap, often species-specific, and easy to apply. But by the turn of the twenty-first century, high costs, undesirable side-effects, and adverse public perception conspired to limit the future of chemical weeding of the range.

Biological control, the deliberate use of a plant's natural enemies, has seemed to many the ideal application of science to range weed control. Commonly species-specific, harmless to other organisms, self-perpetuating, and in theory cheap, its minions range greatly in size—from microscopic pathogens to insects to range livestock. Scientists used insects to more or less control prickly pear cactus in India and Australia, and St. John's wort in the United States. Goats selectively eat brush and other broad-leafed plants worldwide. Some range scientists advocate selective breeding to create livestock breeds that love weedy species. Most biological control projects—and there are many—linger uncertainly in the labs of research and development.

As the twenty-first century opens, most managers concede that none of these methods, singly or in combination, provides the silver bullet that will rid the range of weeds once and for all. Shrub and tree control almost invariably requires eternal vigilance and economic subsidies. More important for the future, all methods require, either directly or indirectly, goodly amounts of that tonic that brought twentieth-century Americans the good life: oil.

The ideal situation for production—grass growing tall and uniformly from fence to fence—remains elusive out on the western range. Many ranchers have resigned themselves to living with woody plants, rugged terrain, variable soils, unpredictable rainfall, and irregular swards of grass. But the urge to standardize, to simplify, seems often just beneath the surface, ready to emerge when control seems possible. Universities continue to spew out scientists trained in the methods of range management and imbued with the hope of grass stretching once again to the far horizon. Meanwhile, the brush marches on.

TWELVE The Last Pariah

Figure 12. Black-tailed prairie dog, Wind Cave National Park, South Dakota. Photo by author.

The turn of the twentieth century to the twenty-first coincided with an occurrence that, though noticed by few, I thought pregnant with millennial symbolism. The incident, which sought to protect an animal capable of shunting a lot of sunshine away from human food production, waved a gauntlet of challenge at utilitarianism, the ancestral philosophy of the ranchman.

The challenge went public on February 4, 2000, appearing as a notice in the *Federal Register.* This daily publication contains rules, proposed rules, and announcements of the federal government. It sometimes reports an instance of taxpayer largesse but also wields the hammer of government regulation. It suggests to many out on the range whether they are in for a smooth ride or about to get bucked off.

This particular announcement sent fear rippling across the prairie like tall grass whipping in a windstorm. It carried the title "Endangered and Threatened Wildlife and Plants: 12-Month Finding for a Petition to List the Black-tailed Prairie Dog as Threatened." The summary below the title sounded like the clap of doom: the U.S. Fish and Wildlife Service had decided that listing this species under the Endangered Species Act was "warranted, but precluded by other higher priority actions." This sounded like the threat it was: We won't bring prairie dogs under federal purview yet, but will if the situation doesn't improve.

Pete Gober, Fish and Wildlife Service biologist and author of the twelve-page notice, carries few marks of a government bureaucrat. Raised in San Angelo, Texas, with a work ethic to match that of an oil-field roughneck, he tackles a job with the intent to finish it quickly and effectively. I've heard him say, only half joking: "Get out of my way or I'll hit you in the head with a pipe wrench." The *Federal Register* notice may not have made him popular back home, but none who know Pete could doubt it came from hard work and conviction.

The fine print, assembled by Pete with help from others in his agency, told the story in detail. The National Wildlife Federation, a respected mainstream conservation group, had submitted the "listing" petition a year and a half earlier. The black-tailed prairie dog, the most abundant of four prairie dog species in the United States, had shrunk in number to perhaps

2 percent of its original population and continued to decline because of plague and poisoning. Several other wild animals of concern to conservationists lived preferentially in prairie dog colonies, among them the highly endangered black-footed ferret, which could not survive without prairie dogs.

But to the ranching community, the whole notion seemed ridiculous. "Just come look on my ranch," some said, "they're everywhere!" "You can't tell me this animal is endangered. It's just another move by those environmentalists to drive us off the land."

Pete agreed that a lot of black-tailed prairie dogs remained—perhaps 800,000 colony-acres from Canada to Mexico. Assuming a reasonable average of twenty animals per acre, that equaled somewhere on the order of sixteen million prairie dogs! Of course, given estimates by early-day naturalist Earnest Thompson Seton that the number before white settlement may have approached five billion, the population indeed had plummeted. This and the continuing steep decline led Pete and others in his agency to look at the letter of the law and decide, yes, this animal warrants listing. Not as endangered—as threatened, which allows more flexibility in management.

But panicked ranchers and sensation-hungry newspapers weren't in a mood to note the fine distinction between "threatened" and "endangered." To them, the Feds' invocation of the Endangered Species Act amounted to a gesture of madness, and that's what mattered. All eleven states in the species' original range up and down the Great Plains and the Southwest responded. Some state people began to devise their own prairie dog conservation plans to encourage the Feds to back away; some convened meetings to plan strategy; some merely frothed at the mouth and shouted.

After the dust had settled, most states agreed to work in a loose organization called the Interstate Black-tailed Prairie Dog Conservation Team. Most also set up their own "prairie dog working groups," which would meet periodically and evaluate progress. With the hammer of the Endangered Species Act in hand, the Fish and Wildlife Service backed off to see what would transpire.

. . . . .

To early-day Anglos crossing the plains, the prairie dog was an interesting curiosity, a distraction from the monotony of prairie travel. Lewis and Clark were so taken with black-tailed prairie dogs that they captured one alive the second year of their journey and sent it back to Jefferson. Josiah Gregg, chronicler of prairie commerce in the decades following Lewis and Clark, described how, when caravans passed, the little animals "frisked about or sat perched at their doors, yelping defiance, to our great amusement." Susan Shelby Magoffin, traveling by wagon train across the plains on the Santa Fe Trail in 1847, showed pleasant surprise when "we came upon 'Dog City' . . . [and] the little fellows like people ran to their doors to see the passing crowd." Southwestern naturalist Edgar Mearns on the New Mexico frontier in the late nineteenth century commented on "the prairie dog as pet, to which every army youngster could contribute something of interest."

That state of blissful ignorance soon would pass. As the 1800s waned, curious travelers gave way to farmers, ranchers, and an array of other entrepreneurs bent on directing prairie production into their own pockets. The federal government responded to pressures from these development interests by creating agencies and ad campaigns geared to production. One agency booster of agriculture, eventually to morph into the U.S. Fish and Wildlife Service, took the name U.S. Biological Survey.

From 1886 to 1910, a man named C. Hart Merriam headed the agency that became the U.S. Biological Survey. From this podium he initiated and directed the first comprehensive natural history surveys in the country. He proved to be somewhat of a maverick in the Department of Agriculture, preferring natural history to resource exploitation, but he paid homage to the agricultural mindset when it seemed prudent for keeping his agency funded.

The name C. Hart Merriam resounds in the hallowed halls of the ecological sciences. At Texas A&M in the early 1960s, I learned of him as the originator of the life-zone concept in community ecology. Two decades later I lived in Flagstaff, Arizona, in the shadow of San Francisco Mountain, where he developed the concept. Hart's Prairie, at the west base of the mountain, now hosts natural history tours and presents a spectacular vignette of the natural human habitat.

In 1901 Merriam published a piece in the Department of Agriculture's annual *Yearbook of Agriculture* that offered up a sacrifice to utilitarianism. Called "The Prairie Dog of the Great Plains," the article showed how quickly the black-tailed prairie dog had changed from frontier curiosity to agricultural pariah. As Merriam put it, the human population had grown, farmers and ranchers had overspread the land, and "the depredations of pests are more keenly felt."

In his yearbook article, Merriam described a "great Texas colony" of prairie dogs that had been surveyed by his most productive field biologist, Vernon Bailey. According to Bailey, the colony covered 25,000 square miles. Pete Gober's hometown of San Angelo grew up at the southern end of this colony, although by Pete's time the colony had dwindled to scattered remnants.

Merriam, assuming that each acre supported about 25 prairie dogs, calculated that this colony alone contained at least 400 million prairie dogs—about twenty-five times as many as the total in existence by the time Pete Gober's notice appeared in the *Federal Register.* Merriam projected the consequences for livestock growers: "According to the formula for determining the relative quantities of food consumed by animals of different sizes . . . 32 prairie dogs consume as much grass as 1 sheep, and 256 prairie dogs as much as 1 cow. On this basis the grass annually eaten by these pests . . . would support 1,562,500 head of cattle. Hence, it is no wonder that the annual loss from prairie dogs is said to range from 50 to 75 per cent of the producing capacity of the land and to aggregate millions of dollars."

Merriam waxed on about the prairie dog's destructiveness: "The damage done by prairie dogs consists in the loss of grass and other crops eaten, or buried under the mounds; in the accidental drainage of irrigation ditches; and in the danger to stock from stumbling in the holes. Running horses often trip and break their legs, and riders are sometimes injured and even killed."

The story got even scarier. Merriam warned that "their colonies have overspread extensive areas previously unoccupied." He repeated testimonials from the field: one colony grew from 10 to 160 acres in ten years, another from two or three burrows to a quarter section in fifteen years. Prairie dog infestation could come to your neighborhood. They are spreading onto private lands from "government, railroad, school, and other lands, over which the inhabitants have no jurisdiction. This is a very serious evil."

Merriam's rhetoric had the calculated effect. It justified funding for his agency to poison prairie dogs, a practice that would solidify into tradition and carry on far beyond Merriam's term at the helm. It also fed a public fear of prairie dogs that still would resonate a hundred years later when Pete Gober, working for Merriam's old agency (now the U.S. Fish and Wildlife Service), proposed to protect them.

Biological Survey reports and other historical records show that Vernon Bailey's "great Texas colony" far exceeded the size of any other colony ever reported. Some say Bailey might have exaggerated, thereby upping the ante for prairie dog control. But he described many other Texas mammals with apparent authenticity. His report of mammal surveys in New Mexico likewise seems factual.

Grazing rather than fabrication may explain the unusual size of the Texas colony. Its location as described by Bailey overlays with uncanny accuracy the last refuge of the southern bison herd. Charles Goodnight himself described the lawnmower effect of bison moving across the land when he arrived in the region with cattle. Large numbers of cattle replaced bison immediately, one mowing machine substituting for another. Without heavy grazing this would have been—and, outside of cultivated ground, still is—mixed-grass country. The grass would have been knee-high to waist-high, too tall for prairie dogs.

Nowhere else in prairie dog range did the transition from intensive bison grazing to intensive cattle grazing take place so abruptly. Elsewhere, time gaps of ten, twenty, or thirty years between the depletion of bison and the massive influx of cattle seemed the norm. Once the bison were gone, as early Kansas naturalist J. R. Mead saw, "Prairie dogs, except a few remnants, disappeared." Then, once cattle came and grazed back the grass, prairie dogs rebounded.

Merriam found it convenient to blame the prairie dog population expansion not on the proliferation of cattle but on increases in prairie dog food and declines in prairie dog predators. Indeed, he might have convinced himself that these were the causes. But then again, his agency needed the political support of stockmen. It might have been unwise to point the finger at them for bringing evil to the range.

• • • • •

In the mid-1990s a woman on a peculiar quest showed up in our community in southwestern New Mexico. She introduced herself as Claudia Oakes. No, she did not have religious tracts to dispense, had no herbal medicines to sell, was not looking for the hot springs. She called upon a couple of biologists and a few old-timers. Could they tell her anything about the prairie dogs in years past?

Claudia's search, as it turned out, did not resemble a random questionnaire—she sought out particular people. John Hubbard topped her list. Recently retired from the New Mexico Department of Game and Fish, John carried the reputation as the best living history book of New Mexico animals smaller than a deer. While with Game and Fish, he'd written a short history of prairie dogs in the state. Interviews with John led Claudia on an ever-branching search for other people, other publications.

Pursuing a Ph.D. degree, Claudia proved unusually persistent in her quest. She dug through journals of early naturalists and pored over land survey records and federal archives in Washington, D.C. She interviewed older ranchers, agency officials, and rodent control specialists, a few almost on their deathbeds. In May 2000, five university professors—her graduate committee—signed off on her 391-page University of Texas dissertation. Entitled *History and Consequence of Keystone Mammal Eradication in the Desert Grasslands: The Arizona Black-tailed Prairie Dog (Cynomys ludovicianus arizonensis),* this tome explains what happened to prairie dogs nationwide in the twentieth century.

Poison was the potion. Gunfire had taken out the last great herds of bison and sent the last Indian to reservation, but it proved inadequate against the burrow-dwelling and prolific prairie dog. Under Merriam's direction, Vernon Bailey and other federal field biologists began testing strychnine, arsenic, carbon bisulphide, and other favorite concoctions of the time. Gradually the method of choice came to be strychnine-laced oats dispensed at burrow entrances.

Military history suggests that people unite most willingly behind leaders when motivated by fear. Adolf Hitler, Franklin Roosevelt, Joseph McCarthy, and George W. Bush brought the multitudes to heel by encouraging paranoia. So did Merriam, and perhaps for much the same reasons—rewards for himself and his bureaucracy. But to be fair to Merriam, the prairie dog campaign eventually grew beyond his control. In the end—as often seems to occur in campaigns that spread fear—he had no power to stop it even if he had wanted to.

Oakes documented in detail the propaganda that vilified prairie dogs and sold their eradication to the public. Others built upon Merriam's success, she said. By the time he left the Biological Survey in 1910, regional administrators and field agents alike had learned to sway opinion and overcome opposition. Cottontails, jackrabbits, and banner-tailed kangaroo rats joined the ranks of evil and suffered the agonies of strychnine; after all, didn't they too eat grass?

Jobs in rodent control multiplied as the first half of the twentieth century gained momentum. With the help of the Civilian Conservation Corps, the area poisoned annually on New Mexico grasslands peaked in Franklin Roosevelt's New Deal years—over a million acres each in 1936 and 1937. Acreages poisoned tapered off in the 1940s. The same general pattern prevailed throughout the Great Plains and the Southwest. The federal government—American taxpayers, that is—by and large footed the bill from the time of Merriam onward.

Claudia sought out and interviewed some of the last surviving prairie dog poisoners. She questioned the people who'd walked the range, back and forth, transect after transect, dropping spoonfuls of oats mixed with strychnine beside holes in the ground, day in and day out. Just making a living.

She asked: What did people in those days think about prairie dogs? The poisoners had different answers, all derogatory. Dangerous to cattle and horses. Pests. Good for target practice. Ate all the grass. What good are they?

Did you ever eat them? No. Couldn't eat anything called a dog. Yes, they taste good. The Indians out here cook them in burlap over a fire.

Wort Youngblood of Truth or Consequences, New Mexico, died soon after Claudia spoke with him. His crew had poisoned prairie dogs out on the Navajo Reservation. The Indian Service said it was okay, he said, but the Indians themselves didn't like it. "The Navajo ladies swept up all our grain and threw it away. Never did get a good kill out there."

. . . . .

In 1899, a different agent of terror arrived in secret on a ship from China that pulled into the San Francisco docks. No one saw the invader come ashore in the dark of night. People found out about it only through its victims, when two sailors on the ship got sick. It turned up in Chinatown the next year, and by 1904 it had killed 118 people.

San Francisco doctors finally learned the identity of the agent. Its strange name, *Yersinia pestis,* did little to alleviate public anxiety. The fear spread when people learned it had wrought far greater devastation centuries before in medieval Europe. Back then people had called it the Black Death, and it had taken out almost half of the European population. Now the California doctors called it bubonic plague. How did it get to California?

Gradually those interested learned the full story. Experts knew *Yersinia pestis,* a bacterium, could spread from animal to animal and from animal to human by fleas. In Europe's plague years, rats and their fleas had found ideal habitat in the dark ghettos of London, Paris, and other concentration camps of the worker class. Since then, though it continued to be known in China and other Asian countries, only occasionally did plague crop up in Europe. It probably got into San Francisco in fleas riding rats that climbed down ship lines to the dock or came ashore in cargo.

By 1908 the plague bacterium had made the leap—presumably in the guts of fleas—from the city rats of San Francisco to wild California ground squirrels nearby. It soon spread to other regions: southern California and Oregon by the 1920s; and Wyoming, Utah, and Arizona by the mid-1930s.

By 1950, plague in wild rodents had crossed the shortgrass country of the Great Plains and entered the mixed-grass regions. Here, at approximately the 102nd meridian of longitude—near the western boundaries of the Dakotas, Nebraska, Kansas, and Oklahoma—it mysteriously stopped. No one knows why. In the half-century since, its eastern limits have fluctuated around this so-called plague line.

Plague has infected at one time or another most of the mammal species that live west of the 102nd meridian. Some species survive plague infections well; others do not. Prairie dogs count among the most sensitive. Colonies of black-tailed prairie dogs hit with plague outbreaks usually lose 98 to 99 percent of their inhabitants.

From Christopher Columbus to George Armstrong Custer, Europeans facing opposition as they colonized the New World found unexpected allies in Old World diseases. Smallpox, influenza, and others jumped from their resistant European hosts to susceptible Native Americans and wreaked havoc. In the same way, plague allied itself with descendants of these European imperialists to help rid the grasslands of a competitor with crops and livestock.

The black-tailed prairie dog found itself caught in a great pincer movement of extinction. From the humid east came farmers, to plow under by the early to mid-1900s the vast majority of the tallgrass and mixed-grass prairies. Most of the acreages not farmed came under "good" range management that favored tall grass and thus excluded prairie dogs. From the arid west during the same period came plague, taking out disease-naive prairie dogs like pins in a bowling alley. The U.S. Fish and Wildlife Service kept its poisoners out on the range, mopping up.

As the twentieth century drew to a close, the federal rodent control people were still out there, still mopping. Only the costs of janitorial staff and brands of disinfectants had changed. Agency administrators still found pest eradication a sure source of funding. C. Hart Merriam might have been astounded at the tenacity of the campaign he'd set in motion a century earlier.

• • • • •

Arizona saw the last of its black-tailed prairie dogs on the ranch of Sandra Day O'Connor. For years I have admired the independence of this first and most venerable woman justice of the U.S. Supreme Court. She has seemed a worthy arbiter of human fairness. But what happened to the prairie dogs on her family ranch?

Naturalist Edgar Mearns gave the first reliable account of the black-tailed prairie dog in its Arizona habitat. He, like later observers, found them in desert grasslands in the southeastern part of the state: "In the year 1885 I observed immense colonies of Arizona prairie-dogs . . . extending as far west as the town of Benson, on the San Pedro River. . . . For miles the burrows of these animals are thickly scattered over the plains south of the Pinaleno Range or Sierra Bonita [Graham Mountains]. . . . Here the 'dogs' fairly reveled and overran the country."

Then came Merriam's rangers. By 1924 they had dispensed poison over more than a million acres of black-tailed prairie dog range in the state. As early as 1932, a young woman naturalist entered a plea in the scientific *Journal of Mammalogy* for preserving the few Arizona colonies that remained. "Control, not extermination," the title ran. Perhaps a man might not have been so bold.

I went to the Day Ranch in 1999. With me rode Tom Waddell, a retired biologist from the Arizona Game and Fish Department. Tom said, "I remember when the prairie dogs were here. They were on public land, part of a BLM grazing allotment attached to the ranch. The rat-chokers took them out." His bitter tone reminded me of his long-standing feud with federal pest control agents.

Tom's pickup truck wound through creosotebush scrub that didn't look much like prairie dog habitat. He stopped occasionally to exit the cab and scan the rolling terrain. Finally he switched off the engine. "It's over there, I think," he said. "Let's walk."

As soon as we reached the brink of the swale, you could see the telltale pockmarks speckled across its bottom. The small depressions, several inches to a foot deep, once had been entrances to prairie dog burrows. After the prairie dogs vacated their burrows and no longer maintained the elevated rims that kept water out, heavy rains coupled with cattle trampling allowed the water to rush in and saturate the clay soil. The burrows collapsed and sealed themselves off, leaving depressions where the entrances had been.

A government survey marker surrounded by stones rose near the center of the now extinct colony. We read the inscription on the brass cap, then oriented ourselves on a map with numbered sections. We stood directly astride the Arizona–New Mexico state line.

"As I recall," Tom said, "the rat-chokers decided to get rid of this town in 1972. They set live traps. Somebody had arranged for the captured prairie dogs to be released on the Appleton Research Ranch down by Elgin, Arizona. You might remember that private ranch, which had been dedicated to conservation and restoration. Some of us in Game and Fish thought it would be a shame to lose the last black-tails in the state, so we suggested moving them there.

"The ones they couldn't catch," Tom said, "they probably poisoned. Anyway, I never saw any here after that. State biologists told me later that the ones they released on the Appleton Ranch disappeared."

On the way back, Tom and I speculated that the Appleton Ranch probably hadn't been the best place for them. It had few burrows dug by other animals that the prairie dogs could use as predator refuges until they made their own. On top of that, grazing had been suspended there, so the grass grew tall. The prairie dogs dumped at Appleton probably just became easy pickings for coyotes and bobcats.

We pondered the meaning of justice, among people and on the range. Who sets the sentence, and on what basis? In 1972, when federal agents came to eliminate the last Arizona prairie dog town, Sandra Day O'Connor was serving in the Arizona State Senate in Phoenix. She probably remained ignorant of the small drama unfolding on her home ranch.

Beginning in the mid-1990s, the Arizona Game and Fish Department escalated its efforts to restore black-tailed prairie dogs to the state. The necessary permissions from the department commissioners and other vested interests proved elusive, however. State biologists told me although the economic and social threats of such a reintroduction seemed minuscule, rangeland dogma still carried great weight with politicians. Reason didn't seem to have much of a chance.

Similar sentiments echoed from the nation's heartland. New waves of fear following the fall of New York's World Trade Towers in 2001 stimulated a general retreat into the fortress of tradition. In 2004 an incumbent U.S. senator from South Dakota fell to a challenger who built a platform partly on prairie dog control. That same year the U.S. Fish and Wildlife Service with little fanfare dropped the black-tailed prairie dog from its list of "candidate" species. At this writing, most of the state prairie dog working groups show signs of crumbling.

• • • • •

On the cusp of the new millennium, our nation seemed ready to wean itself from the traditional All-You-Can-Eat restaurant on Main Street. It sounded ready to cut back, to leave some scraps for the less fortunate. But as new specters of terrorism loomed, it hesitated, looked up and down the street, and contemplated the options. Veggie burgers? Slim Jims? How about skipping lunch? Then it turned, squeezed its hulk through the familiar door it could have found blindfolded, and heaped its plate to overflowing.

THIRTEEN The Trouble with Livestock

Figure 13. Drifting before the Storm by Frederick Remington. Courtesy Yale Collection of Western Americana, Beinecke Rare Book and Manuscript Library.

The project started innocently enough in the next-door yard. My wife noticed a few young men digging holes in the ground with what looked like a post auger, but she couldn't see very well what was happening because a ragged hedge of shrubs separated us from them. Then came the steel pipes.

When I first saw them, they had already been erected. They stood about fifteen feet high, two of them, maybe eight inches in diameter and ten feet apart. Judy explained the construction procedure: The fellows chained the pipes to a backhoe bucket and lifted them, setting the bottom ends into the holes and securing them with concrete.

Then came the crosspiece, a substantial section of I-beam. Welded into place, it spanned the tops of the pipes. A heavy piece of chain looped around its middle and hung down a couple of feet. The whole apparatus looked like the arrangement I'd seen mechanics use to lift large motors from trucks, capable of supporting several tons.

"What is it?" Judy asked the question with some apprehension. She didn't want to ask those who were constructing it, because she didn't know most of them and didn't want to appear totally ignorant of rural ways.

It had a practical application, we suspected. The people who built it—apparently friends of the neighbor's son—almost certainly were not into monument art. Like the neighbor had been until his death a few years earlier, they seemed of ranching stock. A corral of historic vintage abutted the rusty-roofed shed. At unpredictable intervals, cattle arrived in a rattling trailer to bawl and raise dust on the two-acre grounds before they just as unpredictably left the same way.

I guessed it to be the beginning of a repair station for heavy equipment. Despite the residential flavor of our neighborhood, where most houses sit on lots an acre or less in size, little zoning exists and small commercial enterprises come and go. Judy suggested a hanging scaffold to mete out frontier justice.

One morning a few weeks later, a pistol shot boomed out next door. Then another. I went to the window of the building in which I worked and looked out. The reason for the shots merged with the purpose of the tower. Not target practice. Not another OK corral. A cattle butchering station.

A short time later Judy came into my workplace. "They're stringing up the black cow," she wailed. "They're skinning her! Why don't they do it somewhere out of sight?"

The spectacle bothered me less. Having grown up in a family that butchered animals, I saw it more as a clash of cultures than a social outrage. And I had not made friends with the black cow.

A few years later, after Judy's planting of bamboo had grown up to screen the periodic killing of cattle next door, I helped another family butcher a goat. The Coates family seldom ate meat but kept goats for milking. Jim made excellent goat cheese, a type of food I'd seldom tasted before trying his. The pastoral heritage of my own family had come with a definite no-goats code.

One reason the Coateses ate little meat came from their general aversion to killing animals unnecessarily. Yet they also felt compelled to limit the population increase of goats that came inevitably with the production of milk. So they occasionally killed a yearling.

A look of general malaise tainted the family's faces upon my arrival. Eleven-year-old Milagre began right away to recount details of the previous slaughter, which had taken place about a year before. "When it was over I cried," she said with a short laugh.

"We don't have a gun," explained Jim." I killed the last one with a hammer blow to the head. It wasn't a pretty sight—much thrashing about before it finally succumbed. Is there a better way?"

Just in case, I'd brought a .22 caliber pistol with a few low-power "short" cartridges. Jim elected to use the gun. "Sound of a shot won't bother our rancher neighbors," he said. A well-placed bullet brought a quick end to young "Kinati."

Milagre held up well, chattering away in the silence that followed the shot. "Why is so much blood coming from its nose? Do you think it's dead? I'll be saving its skin for a rug." As she rambled on, Jim and I hung the carcass by the hind legs to a low cottonwood limb and began the skinning.

"It's stopped breathing," Milagre said. "Can't feel its heart beating. Can I help skin it? I wonder what the other goats think."

She stretched a long vein she'd freed from the roiling mass that spilled out as we severed the last of the colon from the pelvis. "Feel this! What is it?" She pushed with her forefinger at the stomach, pulled at the stringers of intestinal fat. She had just finished a home-school course in anatomy. Her mother, Nancy, a biologist by training, called her attention to the spongy feel of the lungs and to the heart in its pericardium.

Jim wrapped the carcass in sheets to keep away flies, and we hung it in a ventilated room. He planned to let it age for a week, then barbeque it. Should be really tender by then, we decided, especially since it's only a yearling.

To Milagre's delight, we'd cut no accidental holes in the skin. "I'm tanning it with soap," she announced. "No, I won't be chewing it to make it soft!" As I pulled out of the driveway she, now much quieted, had hung the skin over the yard fence and with a hose was washing the last vestiges of blood from the pale hair of Kinati.

• • • • •

In his book *Landscapes of Fear,* geographer Yi-Fu Tuan wonders whether the rural custom of slaughtering animals might not breed a tendency for violence. He references studies from around the world showing farm folk to be less disturbed than city people by the killing of animals. He points to the likewise greater tendency, at least in America and Europe, for rural folks to kill other people, whether in war, by vigilante lynching, or in support of capital punishment for criminals. The more often people kill, he concluded, the less commonly they show remorse.

Killing animals, suggests anthropologist Rick Potts in his book *Humanity's Descent,* comes with our evolutionary heritage. Philosophers Paul Shepard and José Ortega y Gasset, students of the human hunter, agree. During evolutionary time, the reasoning goes, we killed animals to eat and probably people to defend or expand territory. Though tradition can amplify the drive, genetic heritage equips us to kill for survival. From this perspective, the city people and not the country folk have diverged from the evolutionary directive.

Before being exposed to such philosophical debates, I killed animals to eat in my rural childhood setting. Tree squirrels and rabbits fell to the .22 rifle that came at Christmas shortly after I turned eleven. Before that, I shot robins with my BB gun and whacked armadillos with clubs. Often I heard my family and neighbors criticize meat wastage, but never killing to eat. I did occasionally swear off hunting, as when a dying squirrel would give his last gasps warm in my hand. But the rewards of bringing home meat for the pot soon blotted out the remorse.

On the other hand, killing livestock we'd nurtured since their birth always left me uneasy. The aversion came not from culture, for I cannot remember anyone questioning the slaughter of hand-fed hogs and cattle. It came, I think, from sacrificing those who'd come to depend on me and vice versa. Some of the most glorious times of my youth surround the primal pursuit of wild woods hogs; one of the most poignant came when Old Bully Boy fell to a bullet between the eyes.

Since those days I have found many other people likewise averse to killing animals they've raised. Slaughter of the familiar, the hogs and goats who have become family, raises doubts. Death to strangers seems somehow more acceptable, whether the unnamed victims be cattle in a Great Plains gulag or faraway humans branded enemies or terrorists.

Most of the people who live in my neighborhood, and in the nation at large, eat animal parts disconnected from the animal by the meat-packing industry and a long truck ride. In the wisdom of our civilization, for better or worse, we have eliminated the recipient's confrontation with the donor. We have distanced ourselves from the source of our sustenance.

The very word *livestock* encourages the disconnect. It portrays living things as currency, helps remove the stigma and even the perception that we kill to eat. Most of us seem to be comfortable with this arrangement. Is that good or bad? Who can say. It is the way we are.

• • • • •

The keeping, working, riding, and eating of livestock blends so naturally into human lifeways around the world that most people never think it odd. But in the longer history of animals, our brand of domestication seems pretty unique. Those who study the biological past tell us that life itself started about 3.5 billion years ago, mammals originated about 200 million years ago, near-humans took shape perhaps 1.5 million years in the past, and truly modern humans have been around 200,000 years, more or less. But as far as we can tell, only in the past 10,000 years or so has anything like one species domesticating numerous others appeared. Certainly some ants "herd" aphids, and parasites of necessity receive nurture from their hosts, but those associations seem hard-wired. The widespread domestication of animals by people seems a kind of afterthought independent of evolution, coming upon us unawares.

Domestication of plants and animals arose with the human condition we call civilization. Whether domestication led to civilization, or vice versa, is hard to say. But the two seem inextricably bound together—in time, in place, and in the way we define them. In one sense of the words, civilization *is* domestication—of our plants, our animals, and ourselves.

Historians of animal husbandry emphasize the difference between taming and domestication. As many rural kids know, wild animals often can be tamed—taught to lose their fear of people. They tame best if captured young and raised apart from their species. If brought into captivity before the "critical" age, usually a few days to a week or two, naturally social ones such as wolves, ducks, pigs, and sheep accept people as surrogate parents. At university my advisor studied the behavior of the javelina, that native pig of the American Southwest. He had me raise in succession several taken from their mothers when less than four days old. They quickly adopted my wife and me, our young son, and our dog as herd leaders.

Unlike simple taming, domestication comes through many generations of breeding animals in captivity. Husbandry of domestics almost invariably involves selection for ease of handling and a variety of other traits admired by people. This results over the years in many different "breeds" of dogs, cats, chickens, horses, cattle, goats, and others.

One interesting trait that almost invariably comes with selection in domestic animals is "neoteny"—the retention of juvenile traits and behaviors into adulthood. This not only gives people more control later in the animals' lives, but it also makes the animals more attractive to people. For example, domestic dogs generally show more submissive (juvenile) behaviors and have shorter (thus cuter) faces than wolves, their evolutionary predecessors. My wife's two-year-old cat still sucks on the corner of a blanket when cuddled. Women in particular seem to like pets with neotenous traits.

Indeed, women may have been the primary architects of domestication. I see—and can imagine it happening for millennia—females adopting animals as surrogate children. Although we men usually claim credit for cultural innovations, I think women have the edge in this sphere. They seem better equipped to select animals for tractability and that special look in the eye, to nurture them, and to save at least some of them from being killed when the men get hungry.

Those who study the history of domestication say goats came first. Some archeological digs show goat bones shaped by selective breeding mixed with debris of village peoples as long ago as 14,000–15,000 years. Sheep bones, sometimes hard to distinguish from those of goats, show up at sites of human settlement almost as far back. Reindeer, pigs, horses, cattle, and water buffalo, in about that order, came into human lives as domesticates between 10,000 and 6,000 years ago. Bactrian camels, dromedaries, donkeys, and yaks followed.

All these animals came first into the human fold in Europe, Asia, and North Africa. They arrived almost on cue with other artifacts we associate with civilization. Archeologists and anthropologists say the very first humans to adopt sedentary lifestyles in the Middle East kept goats and sheep, the smallest and cuddliest of the livestock. The bigger animals came later.

In the New World, the list of medium-to-large native animals domesticated contains only two species: the llama and the alpaca of South America, which Andean tribes domesticated between 3,000 and 5,000 years ago. Both these animals descended from the guanaco, so in reality New World people domesticated only one hoofed animal—a short list indeed compared with the dozen or so Old World domesticates.

Except for sheep and goats, the wild progenitors of all the domestic livestock of the world have disappeared, or nearly so. Why? Probably because people hunted them to eat or to remove them as competitors or interbreeders with domestics. Why, in contrast, do goat and sheep ancestors still abound as wild animals? Perhaps because they were smaller, thus offering less return for hunting effort. Perhaps because they occupied remote and rugged terrain hard to exploit by settled peoples.

In any case, for more than ten thousand years some animals have been living so closely with people that they have functioned as family or tribal members. The human-animal association created an ecological "super-organism," better adapted to the hinterlands of human settlement than either human or animal alone. Recognition of the superior qualities of this collaboration may have given rise in early mythologies to such fanciful and powerful hybrids as the man-goat Pan, the man-horse Centaur, and the man-bull Minotaur. The potency of this union continues to mesmerize at least the human component.

• • • • •

Drive west on U.S. Interstate 10 from New Orleans to El Paso, or on I-70 from St. Louis to Denver, and you can see what farmers have known for ten thousand years: the drier the landscape—in this case, the farther west you go—the harder it is to grow crops. Go from South Texas to northern Alberta, or from low to high in the mountains, and you'll see a similar relation between crop production and temperature: the colder the land, the fewer and less productive the farms. Indeed, most of the earth's surface is too dry or too cold to raise the plants we have chosen to domesticate.

In these hinterlands unsuited for crop production, livestock have greatly improved the ability of the land to sustain people. Humans have lived in moderate abundances in the drier and colder lands ever since the first woman tamed the first goat. Abel's sheep carried him into the deserts and mountains far beyond the floodplain farms of Cain. Only on camels could the three wise men have traveled across the desolate dunelands to the city of Bethlehem. The hordes of Genghis Khan from the Asian steppes, the reindeer herders of Europe's far north, the llama tenders of the high Andes, the early cowboys of the American West—only by livestock did these peoples of the outback flourish and multiply.

Livestock have done for people in the marginal lands what domestic plants have done for them in the more productive: funneled sunlight to human advantage. We can't easily digest leaves of grass or dryland shrubs, but goats, sheep, cattle, horses, and camels can. The resulting meat, milk, and mobility have let us thrive in the dry and inhospitable corners of the world. Cain may have slain Abel from his place of power on the farm, but out in the vastness of the grazing lands Abel's tribe carried on, to live by the stomachs of their animals.

As we have seen, all livestock have wild progenitors. Why could people not have been satisfied with harvesting these from the wild? By so doing, they could have avoided the smelly cohabitation and tedious tending that livestock require. They could have forgone the unpleasantness of killing animals they had come to know personally. Why, then, domestication?

The answer seems straightforward: keeping livestock is so much more efficient and less risky than hunting their wild counterparts. Late Stone Age woman had to venture no more than a few steps from the security of the cave to milk or kill a family goat. Abel with a few stones from his sling and a brandishment of his staff kept wolves away from his sheep, and this allowed the herd and the people it fed to grow far beyond their wild potential. Over the millennia, pastoralists multiplied in number thanks to the dependable supply of meat and milk. Like farmers, they eventually outcompeted and replaced those struggling to survive by hunting alone.

• • • • •

Travelers who drive on U.S. Interstate 25 between Las Cruces, New Mexico, and El Paso, Texas, pass mile after mile of spotted cattle and fertile odor. Immense factory dairies line the freeway on the west side—nearer the Rio Grande, which supports a plethora of water-hungry ventures in this arid land. The dairies send out milk and milk products along the arteries of asphalt. Dairy workers dispense untold tons of feed and concoctions of chemicals to elevate the production of milk and minimize that of germs.

In America, the dairy sections of food markets, not to mention many miles, stand between the cows and the customers. Nearly all of us buy dairy products—cheese, yogurt, milk, or at least some of the myriad foods that list milk or milk parts in small writing under "Ingredients." Few link the stuff in their grocery carts to the endless Holsteins that crowd roadside hostels.

In 2006 the Food and Agriculture Organization of the United Nations, generally referred to as FAO, published a 400-page indictment of the modern livestock industry. Said the authors of the FAO document *Livestock's Long Shadow:* "The livestock sector emerges as one of the top two or three most significant contributors to the most serious environmental problems, at every scale from local to global." That despite its estimated contribution of only about 1.5 percent to the global Gross Domestic Product.

How could such a small economic player contribute so much to the world's environmental problems? The world's rural poor depend disproportionately on pastoralism, said the report, and indigent people tend to worry more about the next meal than about the environmental impacts of their livestock. At the other end of the economic scale—the corporation—government agricultural subsidies coupled with cheap energy encourage massive conversion of forested landscapes to ephemeral pasturelands. Efficiency-driven crowding of livestock into feedlots and factory farms concentrates animal wastes that overwhelm the cleansing capacities of nearby soils and water.

More than 95 percent of the sixty million pigs now alive in the United States never see the sunlight, said rancher and lawyer Nicolette Hahn Niman in a March 14, 2007, *New York Times* article. Five million sows inside factory farms endlessly spew out piglets. The sows live out their lives in crates so small they can't turn around, much less avoid the clouds of ammonia and hydrogen sulfide rising from their feces and urine.

Is such concentrated production of livestock good or bad? Producers might say that growing animals indoors protects outdoor environments and increases the efficiency of food production. Population ecologists might question the endless cycle of more food, more people, that results from increasing the efficiency of production. Yi-Fu Tuan might wonder whether our treatment of stock does not spill over into our treatment of each other.

. . . . .

Once, at about the time the twentieth century gave way to the twenty-first, I found myself passing through Dodge City, Kansas. I slowed down, and soon was cruising the back streets of this town that stood tall in my book of American mythologies. I searched about for a glimpse of history.

Surely I could find some image that called up Matt Dillon and the TV series *Gunsmoke*. Its black-and-white lessons of frontier justice had run faithfully every Saturday night in the living room of my childhood home. But I did not find anything resembling the Long Branch Saloon, its proprietor Miss Kitty, the faithful Chester, or any other of Mr. Dillon's friends or backdrops.

I did find a disturbing reminder of the long-past cattle drives from Texas to Dodge. Not a confrontation with pale-eyed gunmen, as in Matt Dillon's time. No, I got lost in a section where brown-eyed stares and Mexican music in the air told me that this was where the immigrants lived. Here huddled families of those who processed meat at the local plants. Here lived the folks who stood between the messy rendering of the cow and the delicate appetite of the average American consumer. As in the glory days of the cattle drives from Texas, Dodge City still served as a trade center for cattle. But Matt Dillon reincarnate might find the present fate of a cow a lot less public and human justice more elusive.

Disconnect describes as well as any word where we have come as a pastoral people. Originally, most herders did their own milking, their own making of cheese and butter, and their own killing. These connections persist in small places and stubborn people. In the urbanized world beyond, where most of us now live, only a tiny fraction of people have seen the bodies go still and the eyes glaze over. Fewer still assume the responsibility for killing their pets. Most prefer their steaks bloodless and nameless behind plastic.

My rural neighbors, in contrast, hear, see, and smell where the meat comes from as they take the beast out with a pistol shot. Intimate contact comes as they hoist the carcass onto a tower or cottonwood limb and gut and dismember it with knife and saw. Some may wince at the death as that of a family member or friend. Some may not.

Unfettered control over nature seems always to bring about ethical dilemmas. So it is with livestock. We have come to shape and protect them, and now we have to kill them. How the protecting and killing should be done will continue to perplex many of us until we get hungry again. Then it will not matter.

In the meantime, we face a hard choice: kill them, or kill the grass. To protect them too assiduously brings us face to face with the specter of Abel's sheep making desert of the Holy Land. It is not easy, being civilized.

FOURTEEN Subsidizing John Wayne

Figure 14. Accoutrements of ranching. Drawing by Casey Landrum.

Most people in the small town where I live depend heavily on a food supply brought from long distances. Even those who harvest backyard gardens, milk goats, or annually kill a deer or cow get most of their calories from town. The food on the supermarket shelves—bread, milk, beans, vegetables, breakfast cereals, bacon, hamburger, and corn and soybeans in a thousand guises—comes from animals and plants grown in faraway places and shipped down highways built and maintained mostly by other people's money. In this sense, we are subsidized.

My dictionary defines subsidy as a "gift or grant to aid the needy." Government subsidies to aid indigent people or elevate corporate profits have become widespread in America, and their dispensation a recurring point of contention among politicians and ordinary people. The federal government provides so-called subsidies to family farmers, ranchers, homeless and jobless people, poor communities, and home districts of powerful politicians. Giant corporations have come to claim more than their fair share of farm subsidies.

Regardless of need, few of those offered subsidies nowadays seem to refuse them. Indeed, the term *needy* in my dictionary's definition seems out of date, having little relation to where government subsidies actually go. As a consequence I have taken to defining subsidy simply as aid from the outside. My community gets a lot of outside aid, sometimes called subsidies but more often not.

Our sparsely populated county qualifies by some government criteria as disadvantaged, but most of us drive late-model cars and only medium-aged trucks. I see no sign that any of us goes hungry—a measure of wealth not many cities can claim—and we don't spend a lot of money on property taxes, water bills, or police protection. We pay only a small fraction of the costs of the roadways, utility lines, flood control structures, and health care facilities that put us more or less on a quality-of-life par with city folks. Taxpayers in other places foot most of these bills.

As politicians well know, careful choice of words can blow a smoke screen over the dispensing of subsidy. Receiving "pork" or "welfare" plays poorly among those who admire self-sufficiency. Getting a "grant" or "contract" may mean the same thing, but it sounds a lot better. I remember my father's outrage at the notion of able-bodied people living on food stamps when job openings existed. But in his later years he picked up free government cheese advertised as surplus commodity. I don't know whether the word *surplus* fooled him, or whether he saw free cheese as small payback for the taxes he'd paid over the years. Regardless, it gets very confusing, trying to figure out who is getting subsidized by whom and whether it's a good or bad thing.

In some ways subsidies resemble addictive drugs. They come in various forms, often disguised as normal ingredients of living. Applied in judicious doses under the right circumstances, some yield benefits. But incautiously applied, they can encourage addiction and generate side effects not anticipated by the provider or recipient. Sometimes the side effects spread to other times and places in unexpected ways.

• • • • •

In the early 1980s I first learned about the impacts of subsidy on Great Basin rangelands. This vast region between the Rocky Mountains on the east and the Sierra Nevada and Cascade Mountains on the west gets little rain and supports primarily ranching. At the University of Nevada in Reno, I called at the office of the man many knew as the eminent range historian of the Great Basin, James A. Young.

Ranching paraphernalia draped the walls of Young's office. Old horseshoes, weathered bridles, bits, branding irons, and faded photographs gave authentic backdrop to his reputation. Here was a man steeped in the lore of the West and educated in the history of ranching.

We talked of the times before cattle, of the silver, gold, grass, and small pockets of water that pulled Americans to the Great Basin in the late 1800s. Of the hopes and dreams of those Euro-Americans who, drawn to these great open spaces, were now replacing the native Americans, so recently diminished by disease and displacement. The grass beckoned deceptively from mountain to mountain. "You'd be interested in a book I'm working on," he said. "I'll send you a draft."

The book came out soon thereafter. I have found no better account of the good times and the bad, the great efforts and the poor judgments, of early-day ranching in the American West. In its pages the ogre of aridity meets the phantom of prosperity. You now can get James Young's *Cattle in the Cold Desert*, co-authored with B. Abbott Sparks, in a new expanded edition from the University of Nevada Press.

According to Young and Sparks, massive changes in the Great Basin rangelands originated in the late 1860s following the American Civil War. Disenchanted ranchers escaping drought in California and postwar reconstruction in Texas gathered cattle, and sometimes sheep, and trailed them to the largely vacant ranges of Nevada, Idaho, and other parts of the Great Basin. They settled around permanent water and started running great droves of livestock "Mexican style"—no fences, year-round grazing concentrated near water, and little concern for stocking levels.

A wide array of subsidies supported the ventures. The original herds themselves had grown on ranges elsewhere, and their numbers in no way reflected the productivity or carrying capacity of Great Basin ranges. The practice of growing hay on irrigated land near water quickly took root, for neither cattle nor sheep came well equipped to survive winters solely on native forage in the cold desert ranges. Boomtowns based on mining of silver and gold—the Comstock Lode and others—sprang up as local markets for meat. The transcontinental railroad, completed across the basin in 1869, offered cheap transportation to markets in California and the East. Banks loaned money liberally for ranching operations. Closely examined, all these benefits turned out to have arisen from an influx of outside capital.

Warning signs began to loom within a couple of decades. In the harsh winter of 1889–90, cattle in many parts of Nevada exhausted hay supplies by midwinter, and by spring a quarter of a million head had starved to death in northern Nevada alone. Work horses, which had multiplied to fill the need for draft animals to grow hay on the 4 percent of the landscape that provided half the annual supply of feed, gave rise to feral herds. These expanded, exacerbating the grazing pressure and pushing grass farther from water. By the end of the century the best of the native grasses were gone, often to be replaced by annual weeds and poor-quality grasses from the Old World.

The native shrubs—sagebrush, rabbitbrush, greasewood, and others—began to expand into the spaces vacated by perennial grasses. These plants, unpalatable at best and toxic to stock at worst, grew inexorably and over the following century turned the former grasslands over vast regions into shrublands. The carrying capacity for grazers plummeted. Today there seems no easy way to return to the pre-cattle condition, to the perennial grasslands that originally lured ranchers into the region.

• • • • •

On the rangelands where I live, the most visible signs of industrial-age man and woman have to do with the capture, transport, and storage of water. To be sure, houses dot the land near towns and cities, and highways cut slices from hillsides and attract the flash and noise of traffic. But out beyond the homes and travel corridors, where cattle outnumber people and much of the land is public, the monuments to technology shape themselves around water—water for livestock.

The kinds of infrastructure differ with the water sources. On the flatter lands, old windmills and new solar pumps bring water to the surface from underground reservoirs and discharge it into steel, masonry, or dirt storage tanks. Higher up, small concrete or masonry basins, most now weathered by time, capture water from mountainside or arroyo-bottom springs. A few old diesel pumps still chug away to pull water from shallow water tables in valley bottoms and push it uphill, sometimes hundreds of vertical feet. Dirt tanks capture runoff from heavy rains.

Ditches and pipes carry the water away from the source. A small concrete ditch captures water from one mountain stream I know and connects it to dirt tanks that tap the ditch for several miles out onto dry mesas and flats. But pipes carry most of the water. In steeper country, gravity flow draws water through old pipes of steel and newer ones of black plastic to otherwise waterless pastures and mesas. Burial kept a lot of the older pipes from sight and freezing, but the new, cheaper plastic resists rupture better and doesn't need to be buried. Near my community, plastic pipes originating at mountain springs crawl down the slopes like black spaghetti.

Conspicuous from highways and back roads are the steel tanks fed by and feeding the pipes. They sit often on ridges and hilltops, storing water from springs or wells and perhaps filling a water trough at their base and one or more tanks farther downslope. They range in appearance from recycled steel containers that in a former life contained jet engines to swimming pool–sized circular tanks of curved steel panels bolted together atop concrete bases.

Symbolic of the water subsidy that spreads grazing to otherwise waterless corners are the softball- to volleyball-sized metal floats in drinking troughs, which close valves to keep water from overflowing. Often at abandoned storage tanks you can find ones no longer in use—rust-resistant, the color of weathered copper, and sometimes chewed by bears. As spreading water across the range prevents grass wastage, so the float valve prevents water wastage. Both together strive toward efficiency in turning grass into cow. For more than a century they have collaborated with barbed wire to try to make ranching in the arid West profitable.

Who paid? On public lands in my area taxpayers generally provided the materials, via the Forest Service, but the grazing permittee often matched the costs by providing labor. I learned this from the Forest Service range conservationist in my district. But as costs of labor have risen in recent years, permittees get proportionately less and less for their dollar. That may explain in part why most of the old steel water tanks I look into stand dry and the cross-fences leak cattle. Indeed, most of the fences, spring developments, steel tanks, and dirt tanks seem to have originated in past times when labor cost less and the federal government handed out more money.

History shows that most of the larger private land holdings in my region, the American Southwest, originated from fortunes made in other parts of the world. The large land grants came first, as gifts from Spanish or Mexican governments. Then, in the late 1800s, wealthy entrepreneurs from elsewhere flocked to the Southwest to grab the biggest and best of what remained. The image of the gentleman cowboy pulled romantics from western Europe, the eastern United States, and other centers of prosperity of the time. As a result, Scottish, Irish, and English names today vie with Spanish ones to dominate historical land ownership rosters and cattle brands.

Then as now, money from the outside often made the difference between success and failure. How can this have gone on for so long without general recognition of the dilemma? John Wayne and the dream of self-reliance cast long shadows. Maybe it was as simple as feeding a desperate hope with comforting words—or as simple as my father's rejecting food stamps but eating "commodity cheese."

. . . . .

Jerry Holechek holds little hope that conventional New Mexico ranching can persevere without more subsidy. For many years a professor at New Mexico State University, Dr. Holechek counts among the eminent authorities on grass and grazing in the American Southwest. His book *Range Management: Principles and Practices,* published in its fifth edition in 2005, educates students and ranchers alike in managing cattle, especially on hot, arid ranges. He laid out his concerns for the future of cowboy culture in a 2001 *Rangelands* article titled "Western Ranching at the Crossroads.".

Declining profitability heads the list of problems, says Holechek. The price of beef, adjusted for inflation, has declined substantially since 1980. When coupled with escalating costs of producing beef—brush control, transportation, meeting new environmental regulations, declines in range carrying capacity—ranchers face a bleak future unless they figure out ways to bring in money from other ventures. Among these he suggests fee-based hunting and fishing, wildlife viewing, getaway lodging, golfing, trail trips, and rodeo stock production.

Eight years earlier, Holechek introduced the urgency of the situation with an economic analysis in *Rangelands* called "Desert and Prairie Ranching Profitability." He saw western cattle ranching as "definitely a high risk–low reward proposition." Operations costs—insurance, utilities, taxes, veterinary care, supplemental feeding, state land grazing fees—increased at an overall rate of 5–7 percent per year in the 1980s, he said. Many arid-lands ranches soon might be forced out of business unless they could bring in outside money. An unrelenting problem is competition with regions that get more rain and sometimes cheaper labor—the eastern United States, Argentina, Brazil.

One of the best-kept secrets of western ranching's success in the twentieth century turns out to have been oil. It powered ranch trucks from the 1930s on, as well as bulldozers to build dirt tanks and ranch roads and to uproot invading brush. For decades it has run the eighteen-wheelers taking cattle to market and built and maintained the highways on which they traveled. It has given rise to fertilizers and pesticides for growing hay and grain supplement, and has fueled the irrigation pumps that made millions of desert acres "bloom like a rose."

A name that shows up frequently in books and articles that lead scientific thinking in the field of range livestock production is that of Rodney Heitschmidt. As chief editor of the book *Grazing Management: An Ecological Perspective,* he assembled and edited technical papers by an astute group of scientists. Their expertise covered all the important topics—soils, grass production, livestock foraging behavior, nutrition, and social and economic influences.

In 1996 Heitschmidt and two other researchers published an article in the *Journal of Animal Science* titled "Ecosystems, Sustainability, and Animal Agriculture." In it they sought a way to expose the hidden subsidies in livestock operations so as to judge long-term sustainability. Their method of analysis, using energy output/input ratios, had been used for some time by ecologists to describe how ecosystems function.

The idea is simple: Energy, measured in food as calories, governs production in plants and animals. Energy from sunlight builds grass, and energy from grass supports cows. For an agricultural production system to be sustainable, the amount of energy coming in (inputs) always must be greater than that going out (outputs, or production). The "yields" of agricultural systems thus are governed by the abilities of plants to fix energy from sunlight and the abilities of humans to efficiently harvest the products. Because increases in available water elevate the abilities of plants to capture energy and make it easier for people to manage livestock, wet regions show more favorable output/input ratios and have higher yield potential than arid ones.

Heitschmidt and his colleagues calculated energy inputs and energy outputs for eleven different kinds of beef management systems in eastern Montana. They came to a sobering conclusion: typical livestock operations in the northern Great Plains can be sustained only by continued heavy inputs of energy from fossil fuels.

From the southern ranges came agreement by Jerry Holechek. In a 2006 *Rangelands* article, "Changing Western Landscapes, Debt, and Oil: A Perspective," he predicted that coming scarcities of oil would drive up the price of ranching. However, he offered the struggling rangeland cowboy a buoy of hope: as oil prices rise, your beef may compete better than that produced by feedlots, which depend even more on fossil fuels than you do.

Bottom line: costs of beef production by any method will escalate everywhere unless we come up with a good, cheap substitute for oil. The prospects for that look slim indeed, said Professor Holechek. In his 2006 article he predicted what came to pass in 2008: oil topping $100 a barrel. Looming recession and lower oil demand have since reduced the cost, but some respected analysts predict $200 a barrel within a few decades.

Will future beef-eaters in Los Angeles and New York be able, and willing, to pay a lot more for their steaks? It seems doubtful if the current economic decline continues. Beans will look better and better. Keeping John Wayne alive in the arid lands may require cutting cattle from the script, except as bucking bulls in rodeos or trail steers driven by wealthy city slickers.

FIFTEEN Collateral Damage

Figure 15. Dangerous habits. Drawing by Casey Landrum.

Mushroom cloud! It rose ever higher into the clear blue sky. The top billowed out like a cauliflower growing at time-lapse speed. A slice of horizontal gray separated the cap from the stem. A low rumble of thunder sent a shiver across the glass window of our old house.

But this couldn't be a thunderstorm. The calendar proclaimed it to be too early for the monsoons of summer, and the air felt bone dry. But in yonder sky, the roil of white pushed higher, many thousands of feet into the air, seeming to defy the laws of meteorology. The sinister band of gray expanded. And thunder rolled.

The previous winter of 2005–6 had ended as the driest on written record in New Mexico. I had never seen anything to match it since first coming to the Southwest forty years earlier, and old-timers with twice that time in this land said the same. Winter in New Mexico and Arizona usually brings Pacific moisture, but not that year.

Spring took up the cadence where winter had left it. Time marched on, bringing week after week of cloudless sky. But spring seldom sees much rain here anyway, and we didn't anticipate a lot of moisture until July.

By mid-June, many southwestern regions had endured eight months with less than an inch of rain. During the days the temperatures now climbed toward triple digits and the relative humidity readings withdrew to single ones. Grass not grazed stood yellow and tall from a lot of rain the previous summer. It rippled ominously in the breeze.

June 20 dawned another sky-blue day. That morning I had noticed a few wisps of smoke on the horizon to the northeast. At about two o'clock in the afternoon, while tending to a few outside chores, I felt a pulse of pressure as from distant dynamite and looked up. The apparition loomed.

An unprintable response came to my lips. Images of atomic blasts had become commonplace of late in political rhetoric and on media screens. Here one rose up in real life, high above the crest of the Mogollon Mountains.

A neighbor driving home that day from the west thought the cloud might have arisen from a giant explosion set off by the military. Having heard rumors that a test blast involving megatons of ammonium nitrate and fuel oil had been scheduled for the White Sands Missile Range some hundred miles to the east of us, he wondered if this wasn't its aftermath rising to the heavens.

It wasn't. It was the Bear Fire blowing up. Those familiar with wildfire say such explosive ignitions occur when fires encounter especially hot, dry, and fuel-rich conditions. Lightning and thunder come from friction generated by the rapid mixing of the particles in smoke. Starting near the fire lookout tower on Bear Wallow Mountain and sending the attendant Mary Ann racing for safety, the Bear Fire registers among the most spectacular natural events I have witnessed. And this from ten miles away.

The unusually thick and tall dry grass—called "fine fuels" by the fire people—had waited for a spark to set it off that early summer of 2006. The fine fuels ignite easily. In years when they are abundant they can carry fire to, and ignite, the coarser fuels—the shrubs, trees, and patches of woody debris that stand too far apart to carry a fire by themselves. Abundant fine fuels can turn an ordinarily fire-resistant landscape into a tinderbox.

Anticipating trouble, the U.S. Forest Service had issued advance contracts to the famous Navajo firefighters. Fire-support contractors alerted the unemployed on their lists, hoping to avoid the inexperienced and the sensation seekers. Agency "hotshot" crews rode around in green buses. Forest Service people watched from mountaintop towers and computer stations. Business picked up at small-town restaurants.

Within two weeks after it started, the Bear Fire had consumed some fifty thousand acres—nearly eighty square miles—of scrub oak, pinyon, juniper, ponderosa pine, fir, and spruce. Initially the dry grass had carried the fire from shrub to shrub, thicket to thicket. Once hot enough, the fire had crowned in the denser stands of pine and fir, leaping from tree to tree.

Other fires erupted that June, though none in our region to match the Bear Fire. Smoke dimmed the mountains during the day, choked the valleys at night. Along roads and in villages I saw firefighters from Oregon and Arizona. Large print on the sides of their buses advertised their pride and their origins.

The Bear Fire came under control by early July. To those of us inexperienced in the business of controlling fire, the early monsoon seemed more effective than either the crews trying to contain the fire on the ground or the tons of slurry dropped from the air. But what if the rains had come late? How soon could the ground and aerial attacks by firefighters have contained the fire? As it was, a few remote summer homes along Indian Creek near Bear Wallow Mountain had burned. Would the losses have spread to the many more homes along Willow Creek?

Once the Bear Fire had died to wisps of smoke from smoldering logs, the aftermath dominated conversations in local restaurants and other meeting places. "Used to be a beautiful place," one person said; "I hate to drive up now and look." "A part of me has died," said another who had worked the area for many years as a Forest Service employee. "A catastrophe," most agreed.

A glorious profession it is, fighting fires. In my community few complain at the costs, spread out as they are among taxpayers across the country. We send our sons, daughters, husbands, and wives to share in the spoils.

. . . . .

In late fall, five months after the fire, a friend and I took Bursum Road up the mountain to take a look. A half hour east of the village of Mogollon, near the summer community of Willow Creek, we began to encounter scorched patches of ponderosa pine at about the 8,000-foot level. An understory of what looked like domestic ryegrass, which had been seeded from the air following the fire, hung heavy with seeds. As we drove toward the burn's interior, the blackened patches enlarged and coalesced into general devastation. The seeded grass showed the only green.

A mile beyond Willow Creek, the road ended. Previously a well-used thoroughfare that went on to Snow Lake and the county seat of Reserve, it now vanished where it encountered Gilita Creek. Uprooted trees and boulders up to auto-size hid what roadbed remained; the rest had disappeared somewhere downstream. The tops of concrete picnic tables at the nearby Gilita Campground projected only a foot or so above the sand, gravel, and cobbles brought down by the flood. The deposits of sand on the privy floor raised your knees to your chin.

In local jargon, the creek had blown out. Looking beneath the standing black stick-trees and the green floor of rye, you could see why. Water from the rains following the fire had cut hillside rivulets, which deepened into channels, which spewed out sand and gravel to make alluvial fans at the foots of slopes. Multiplied manifold along Gilita Creek, such accelerated runoff had generated a flood level and a rearrangement of sediment not seen in the history of Bursum Road.

Record rains had amplified the damage to be sure. Yet unburned areas nearby showed little in the way of abnormal erosion. There, the tree canopies had intercepted the rain, the accumulations of conifer needles and duff had absorbed the water and slowed the runoff, and in the wider openings native grasses had spread out the water and held the soil.

Where the fire had raged, however, all these checks were eliminated. Those driving to the end of the road marveled at the scoured creekbed, the accumulations of debris, and the underground telephone cable and metal culverts unearthed where the road had been. Some even endured the discomfort of the privy.

A lot of elk and deer tracks laced the new gravel bars along the creek. The animals had come not to view the damage but to feed on the seedings of rye. Bear droppings showed an unusual tinge of green—probably they also had discovered the seedheads, hanging plump at bear-mouth level. Some years might pass before the deep interior of the burn would attract a lot of animals, but the large mammals at least seemed less prone than we to judge it catastrophic. Animals have a tendency to go where the food is.

. . . . .

The Forest Service blamed the Bear Fire on a careless camper, as yet unidentified. But like so many other events we call disaster, the calamity of the Bear Fire originated less from individual carelessness than from entrenched philosophy. Some would say the responsible ideology in the Forest Service arose in part from a man named Gifford Pinchot.

Born in 1865 into a family made wealthy by lumbering, Gifford Bryce Pinchot dedicated his life to righting some of the wrongs inflicted by a cut-down-and-get-out philosophy. Following a stint in France studying forestry (and leveraging his father's money), in 1900 he and a Yale classmate founded the Yale School of Forestry—the first such educational venue in the United States. Thanks to family connections, he rose to political prominence under the patronage of President Theodore Roosevelt. While holding the position of professor at Yale from 1903 to1936, in 1905 he became the first chief of the newly created U.S. Forest Service.

In reaction against the exploitive practices of the day, Pinchot condemned waste. He instituted economic efficiency and sustainable yields as goals of the new agency, calling on science to show the way. To disarm critics who thought government should not be meddling in timber management, he vowed to make forestry profitable on the millions of acres so recently brought under government control by his friend Roosevelt.

We now recognize Gifford Pinchot as an early leader in managing renewable resources. It was he who first applied the term *conservation* to sustainable use. My professors at university called him the father of American forestry.

As Forest Service chief Pinchot toured the vast ponderosa pine timberlands in the highlands of Arizona and New Mexico. He thought they could be managed better. In his book *Breaking New Ground,* written shortly before his death in 1946, he described looking across a woodland prairie at "an Apache getting ready to hunt deer. And he was setting the woods on fire because a hunter has a better chance under cover of smoke. It was primeval but not according to the rules."

Pinchot felt a compulsion to control this wild Promethean way. At that time, the southwestern forests looked more like savannas than woodlands, with trees widely spaced and with a dense carpet of grass in between. Indeed, Pinchot's Apache probably set fire to the grass rather than the forest. In earlier times at any one site, fires that burned the grass recurred about every five to ten years, usually set by lightning but often by Indians. To Pinchot, the travesty was that the fires killed nearly all the tree seedlings. The rules of forestry, after all, called for growing more trees.

To Pinchot's good fortune, the Apache "menace" to white settlers had subsided a couple of decades before his ascendancy to head of the Forest Service. In consequence, sheep and cattle had multiplied in what soon became the national forests. Although sheep ate some of the pine seedlings, grazing generally proved an unexpected ally, for it removed the fine fuels that carried fire. To augment this luck, Pinchot instructed his men to put out fires that managed to find enough fuel to burn.

Fire frequency quickly dropped, to one-tenth or less than it had been before. In years with suitable moisture, many millions of tree seedlings took root, multiplied, and grew. The year 1919 proved a particularly good seed-germination year for ponderosa pines in Arizona and New Mexico. The open savannas began to close.

The unanticipated side effects of the new forestry came slowly and, to many, imperceptibly. Before grazing, the fires had burned cool because little fuel built up between recurrences. Though usually hot enough to kill seedlings, these fires could neither ignite nor kill the larger trees. The perennial grasses, likewise fire resistant, proliferated. They slowed water runoff and minimized erosion. Floods peaked at modest levels, and trout habitat persisted. Roosevelt himself came to the magnificent hunting grounds of Arizona's Kaibab Forest, where deer moved through the open forests in large droves.

Enter livestock and the Pinchot philosophy. Tree seedlings survived and began to overshadow and choke out the grass. The water infiltration that fed local aquifers and springs subsided as tree canopy expanded and grass cover declined. The load of coarse fuels, both living and dead, escalated. Young trees clustered around the big trees that produced them and began to take on the threatening shape of "fire ladders" that could carry a cool ground fire into a hot crown fire. Cattle, deer, trout, and a host of other animals declined in number.

In June 2006 the Bear Fire blew up. Flames quickly climbed to tree crowns and the temperature skyrocketed. We in the valley gawked at the mushroom cloud. Soon it was over, and a sea of black sticks stood stark against the mountainsides and the sky. The rains came, and Gilita Creek ripped down its valley with a roar of boulders and a toppling of trees. And the Forest Service sought a living perpetrator.

• • • • •

Given the Forest Service's dedication to management by science, how has the policy of fire suppression persisted so long on public lands? Warnings came soon enough. Shortly after Pinchot initiated the policy, one of Yale's own recent graduates published a scathing commentary on the evils of fire suppression in southwestern grasslands. The young forester's name was Aldo Leopold.

Leopold began coursework at the Yale Forestry School in 1906, six years after the program took shape and a year after its founder and professor,

Gifford Pinchot, took charge of the new federal Forest Service. Soon after Leopold graduated with a master of forestry degree in 1909, he took his first assignment—as forester of the Apache National Forest in eastern Arizona—not far from where the Bear Fire would erupt a century later.

As a green young scientist, Leopold began to steep himself in the lore and lifestyle of the southwestern high country. He competed with the cowboys to fight the snow and be the first to "top out" on the mountains in the spring. He pursued the predators and chronicled their demise—the last grizzly of Escudilla, the dying fire that was the southwestern wolf. Trained to observe, he did so with an eye resistant to the dark glass of dogma.

Leopold's article, "Grass, Brush, Timber, and Fire in Southern Arizona," came out in 1924, in the forestry profession's official chronicle of science, the *Journal of Forestry*. In it he summed up his fifteen years or so of riding the backcountry, talking to ranchers, looking for patterns on the land, and thinking about conservation.

His story begins: "One of the first things a forester hears when he begins to travel among the cow-camps of the southern Arizona foothills is the story of how the brush has 'taken the country.'" At first skeptical, Leopold began to see for himself the signs on the land: the fire scars on the older trees but lack of recent fires, the diminishment of grass cover, the profusion of tree seedlings moving into grassy openings, the widespread abnormal erosion. On close examination he dated the beginnings of the invasion at about forty years previous—the time when the first livestock came in force.

For the past fifteen years, he said, "we have administered the southwestern forests on the assumption that while overgrazing was bad, fire was worse." Then he seemed to take a shot at what he learned at Yale: "While there can be no doubt about the enormous value of European traditions to American forestry . . . there can also be no doubt about the great danger of European traditions uncritically accepted and applied."

The Leopold Vista Historical Monument sits beside U.S. Highway 180 ten miles south of Glenwood, New Mexico. From there you get a good view of the Mogollon Mountains to the east. In 1924, the same year he wrote about the hazards of fire suppression, Leopold helped designate part of the Mogollons as the nation's first official wilderness: the Gila.

Today from Leopold Vista, you can see in the near distance young juniper and mesquite trees in the slow-motion act of swallowing grassland. Casual travelers picnicking at the rest stop may have neither the interest nor the skills to translate the young ages of the woody plants into future loss of grass. Following wet summers the highway department people mow the grass along the nearby highway, in part to keep these travelers' cast-off cigarette butts from starting range fires.

Leopold took an unusual interest in the animals of the forest. In time he would lead the growing movement to conserve habitat for animals other than human. Many in my profession call him the father of wildlife management.

. . . . .

So why, more than eighty years after they should have known better, does the Forest Service still put out wildfires? Jobs, infrastructure, and habits of thought have grown up around fire suppression. Homes of people have proliferated on the grasslands and in the forests, partly because fire suppression seemingly has reduced the risks, and demands for yet more fire protection rise accordingly. Coarse fuels have accumulated to the extent that some forest supervisors prefer to postpone the inevitable infernos as long as possible—ideally until they retire or move away and can hand the problem down to supervisors who come after. And perhaps, in most of us, there still resides an evolutionary fear of fire shaped over hundreds of thousands of years of living in habitat shaped by fire.

The collateral damage of fire suppression ranges far beyond the southwestern grasslands and the U.S. Forest Service. It lures California homeowners into the very heartlands of historic wildfires. It burdens rural communities throughout the American West more and more with the escalating costs of fire protection. It pulls a host of woody competitors into grasslands—eastern redcedar into tallgrass prairie, mesquite and other shrubs into the southern Great Plains and desert grasslands, sagebrush into Great Basin pastures, aspens and conifers into mountain and subarctic meadows.

Fire suppression has turned the slope of the mountain I see daily out the front window into a brushland where only the black bear and the rufous-sided towhee hold forth. The further thickening of the shrubs and trees threatens to reduce the habitat quality for even these animals. At the local level, fire suppression promises to be among the last nails in the coffin of the working American cowboy. Worldwide it inexorably compromises our natural habitat and that of our favorite animals.

• • • • •

Savanna is an ecosystem in which grass provides the fuel that ensures its own survival, says botanist W. D. Clayton. In a 1981 article titled "Evolution and Distribution of Grasses," this British botanist described how savanna environments around the world evolved under fire. Other experts echo Clayton's thesis. Indeed, few seem in serious disagreement. What this means is that recurring fires over the fifty million years or so since grasses took shape on earth maintained the environment that gave distinctive shape to humans. We are creatures arisen from the crucible of fire.

During most of our time on earth we may have assisted lightning in the ignition of our habitat. Some experts say our species has existed perhaps half a million years. By 460,000 years ago, infers anthropologist and archaeologist Rick Potts in his book *Humanity's Descent,* our predecessors seemed to have gained control of fire to some extent. The evidence is there on ancient cooking hearths and in fire-blackened caves.

Probably also by that time our ancestors had taken to firing the grass at will, though here the evidence is less certain. Undoubtedly those ancients could see that burning the grass exposed the predators that would eat them, destroyed the pests that would bite them, and benefited the grazing animals they would eat. Thus by fire did our kind begin to flex its evolutionary muscle, to seriously modify its own habitat, and perhaps by this path to venture further onto the treacherous ground of hubris.

Only in recent history, it seems, have we taken on the role of wildfire suppressor. People always must have observed that it's a lot easier to start a fire than to stop one. Suppressing fire goes against the natural tendency for fine fuels to burn, and thus energetically saps the enterprise. As with modern livestock production, the real cost of fire suppression in the twentieth century hid itself behind cheap energy from fossil fuels. Thus as oil prices continue to rise and competing demands for it to multiply, the glory of hotshot ground crews and aerial slurry runs may well fade from our culture.

. . . . .

In the mid-1990s, the Saliz Fire some twenty miles north of our town burned several hundred acres of rugged rangeland. Historical accounts of the surrounding range and a close look at the ages of the woody plants nearby suggested that a century ago the site supported mostly grass. But as in many other areas at similar elevations, juniper, pinyon, scrub oak, and other brushy species have proliferated during the intervening decades, eliminating most of the grass.

A curious change in animal use followed the fire. Because the burn bordered Highway 180, which I traveled periodically, I stopped on occasion to look for plant regrowth and animal sign. Hunters, ranchers, and field biologists spend a lot of time looking for sign—tracks, droppings, bedgrounds, and other disturbances you can see just by walking around. Animal sign gives you a lot of insight with a minimum of observation time.

As at the Bear Fire, nearly all aboveground parts of trees and shrubs had been killed. But within a few months post-fire, the fire-tolerant shrubs had resprouted from the root crowns. Alligator juniper, gray oak, manzanita, and skunkbush sumac advertised their survival by thrusting up shoots from their charred stumps. A maze of animal tracks led into the burn, and fresh piles of deer and elk droppings littered the ground. A closer look showed why: the animals had hedged the manzanita, sumac, and other newly emerged plants inside, but not outside, the burn.

Vegetation that sprouts after burns almost invariably attracts deer, elk, cattle, bison, horses, and other herbivores. Analysis of plant tissue explains why. The new growth—whether of grass, weeds, shrubs, or trees—is more digestible and nutritious than that of the same species outside the burns. Animals come to where the food is best.

I have looked at aftermaths of a lot of wildfires in rangelands diversified by topography. The fires themselves tend to vary a lot in temperature, proportions of aboveground plant growth destroyed, post-fire regrowth, acceleration of runoff, and soil erosion. Most burn nonuniformly, leaving behind a patchwork of unburned, lightly burned, and heavily burned sites. Animals generally like the patchwork effect. Being interested in animal diversity and abundance, I have appreciated what most fires did to the land. In my eyes, only a few of the really hot ones did damage that outweighed the long-term benefits.

I suspect that, like me, most people judge a fire good or bad depending on what they value in the landscape. Those interested in an abundance of deer, elk, bears, grouse, and quail usually appreciate the rejuvenation of these animals' habitats by fire. People who respond mainly to the postcard green of forests tend to despair at the black sticks of a crown fire's aftermath. The changes almost always will benefit some kinds of animals and repel others. But few people in fire country like to think it is only their particular set of values that makes a fire good or bad.

In time, naturally recurring fires will come again into their own. Such burns often will be ragged and irregular, and sometimes spectacular to behold. With increasing fire frequency, fire intensity will diminish and grasses will begin to expand, a patch here, a sward there. Brushy thickets and choked forests will dwindle. Patchwork in vegetation cover will intensify. Sometimes streambeds will blow out, then rebuild, perhaps multiple times. Some folks—I hope many—will applaud the return of our evolutionary ally and with it the increasing variety of animals outside domestication.

Unless we by some miracle maintain or increase our per capita wealth, fire-suppression subsidies to the hinterlands will inevitably diminish. Money and people from the outside will intervene less and less into the vagaries of nature. Soon large wildfires will become only small blips on a national consciousness increasingly preoccupied by inundations of coastal cities, international struggles over oil, and perhaps worse. Once again we will adjust our lives to the supremacy of fire.

SIXTEEN Cowboy

Figure 16. Arizona cowboy. Photo by Patrick O'Brien.

I first got drawn into the glory of the American cowboy by Zane Grey. His books caught my attention in about the fifth grade. This Pennsylvania dentist turned western novelist spun tales of pure women and rugged men who by and large followed the rules of Victorian social etiquette, and his stories kept me huddled under the bedside lamp often until the small hours of morning. Once I had read all the Zane Grey books owned by our school and town libraries, I began to reread them. One of the few possessions of another I really envied was a matched set of Zane Grey owned—but, beyond belief, not read—by a school classmate.

An antique copy of Grey's *Under the Tonto Rim* still sits on my bookshelf. It, together with a few short stories—assembled in Zane Grey's *Western Magazine,* volume 4, number 5—hides yellow and brittle between replacement covers of cardboard I reinforced a half century ago by gluing on feed-sack cloth featuring horse heads and riding paraphernalia. The smell of its pages rekindles the days of youth, recalls the distant prospect of cowboy country.

Long after I first encountered cowboy novels, I learned that Owen Wister might have started the genre in 1905 when he published *The Virginian.* Like so many pulp westerns to follow, it featured the chivalrous horseman, the chaste woman, and the evil antagonist, set against the far horizons and close barrooms of that small window of historical time that was the late 1800s. In time I also found Max Brand, Luke Short, Louis L'Amour, and a number of other well-known writers who followed the same script. But in my mind, none quite measured up to Grey.

It has been many years since I have read Zane Grey or any of the others. Nonfiction has pulled me away, and I suppose most western novels would seem a little contrived, now that I have traveled much of their settings. But in retrospect, I still marvel at their power to call so many Americans in my age cohort at least figuratively into the west, that far-flung grassland of fabled self-sufficiency, big animals under control, and women on pedestals.

Moviemaker John Ford amplified the power of the western novel by showing the scenery. By shrewd calculation he brought the habitat to movie houses and drive-in theaters across the country. Several lesser movie directors also helped bring the cowboy to the screen, but Ford exploited the innate appeal of sweeping landscapes. He eventually settled on Monument Valley in southern Utah as his movie set; there, where the red bluffs rose up and the horizons looked limitless, he made his name.

With an artist's instinct Ford discovered the power of vertical relief and the far view. If you have driven through Monument Valley in the early morning or late afternoon when the shadows reach far and have not been moved, your ancestry surely has drifted apart from the human lineage. It is our natural habitat, supersized. Into this landscape rode a bigger-than-life protagonist who would star in more than twenty of Ford's films, a young man named John Wayne.

As the western range itself had done, the "Duke" gained admirers around the globe. A group of Canadian colleagues I once brought to the Southwest gloried in the summer sizzle among the saguaro cacti as they relived the fantasy of Wayne confronting the desert. A German visitor to my town in New Mexico looked forward with anticipation to visiting nearby Lordsburg out on the flats—to see the supposed setting for Ford's and Wayne's early movie *Stagecoach,* based on Earnest Haycox's novel *Stage to Lordsburg.* It turns out that Ford made the movie not around Lordsburg but in Monument Valley far to the northwest.

The western horseman rode tall from the beginning. Soon after pulp westerns began to capture American readers, Jack Culley, Oxford-educated manager of the huge Bell Ranch in eastern New Mexico, judged that the "popularity of the cow business over the sheep business in modern literature is primarily due to the fact that the cowboy carries on his business on horseback, whereas the sheepherder goes afoot. So in old times the knight and his charger got all the press notices and the favors from the ladies, while the pikemen, who operated on foot, remained unnoticed."

Horses make giants of men. Crusading Christians, Mongol hordes, and desert-dwelling Arabs in the Old World, and Spanish and English *conquistadores* in the New, collaborated with horses to overcome superior numbers. The originally foot-bound Comanche, struggling to survive in the foothills of the Rockies, leaped to domination of the southern Plains on the back of the horse. Only with horses, said historians Walter Prescott Webb in *The Great Plains* and Terry Jordan in *Trails to Texas,* could the cattle industry and cowboys have achieved such stature in the American imagination.

My neighbor Kenneth Shellhorn, who is dead now, used to reminisce about working cattle and hunting cougars. He had spent his entire life in this country, he said, and preferred mules over horses—but either beat going afoot. When introducing me to another horseman, he would sometimes allude to my aversion to riding—my handicap—with a brief "He walks."

The horse's bulk not only gives the rider a material advantage over pedestrians, but it intimidates as well. Many of our fellow mammals—from two-pound dogs to five-ton elephants—try to intimidate each other by bristling, standing sideways, stretching tall, or in other ways looking as big as they can. Some anthropologists guess that the tendency for our own hair to rise with fright comes straight from the Pleistocene, when hair standing on end probably forestalled risky fights by making the fearful look bigger.

Indeed, the American cowboy in traditional garb looms large even when dismounted. Cowboys big and small can look imposing in high-heeled boots, big hats, and big pickups. They claim the psychological high ground when confronting hatless city folk wearing sandals and shorts and driving fuel-efficient sedans. Is the tendency for some people to wear Stetsons even on airplanes and Dallas streets a bid for domination, an advertisement of tribal affiliation, or just habit? I guess it probably depends on the individual.

An anthropologist might say that women also pay attention to size, but in a different way. In ancient times the strength that came with a man's size, and the status it conferred on his family, might have meant the difference between life and death of the woman and her children. Today, wealth and power may hide behind more subtle signals, and certainly I cannot speak to what attracts modern women. But over the years it has seemed to me that the cowboys with the alpha accoutrements—even if destitute from chasing rodeos or Saturday night poker—get more than their fair share of feminine attention.

• • • • •

The ranching lifeway continues to mesmerize many Americans. To begin to understand the endurance of this affection, let's take a look at the monetary returns. Part of the attraction of pastoralism in the West was, and is, the idea of economic self-sufficiency. But does it make economic sense to ranch?

In the middle 1880s, Captain William French took over the management of the WS Ranch in southwestern New Mexico. Still called the WS and located a few miles down the road from where I live, in French's time it had recently been bought by two wealthy Britishers. Captain French, himself an Englishman and recently arrived in America, seemed also to have money. He later published his experiences in a book, *Recollections of a Western Ranchman.*

In 1887, French assembled the first trainload of WS steers—fat four- and five-year-olds—and trailed them northeastward to the railhead in Magdalena. (Today, west of Magdalena beside U.S. Highway 60, a roadside plaque commemorates the trail they probably took. In Magdalena itself you can see what's left of the old livestock holding corrals.) Once loaded onto the train, the WS cattle endured the long ride east to Kansas City, a center of cattle commerce at the time. Upon their arrival, they brought three dollars per hundredweight, a nice price in those days. The future looked good.

Quickly French had the WS outfit gather another herd, all in good shape and, French thought, maybe worth three dollars and ten cents. But by the time the second load reached Kansas, the market had started down. Telegrams flew back and forth, and after holding out for a rise in price, French decided to overwinter the WS cows on Kansas range. The herd eventually sold the next year, at less than a dollar a hundredweight—forty cents per hundredweight less than the expense of getting them there, French calculated.

Such has been the western ranching business from its inception onward. Walter Prescott Webb in his masterwork *The Great Plains* tracked the ups and downs in the industry from the Civil War to 1930. *Volatility* is the word that comes to mind. The unpredictability of blizzard, drought, and market prices built a bronco-busting industry in which the occasional high rides spawned hopes that more often than not tumbled into the dust.

Water mattered. The drier the land, the more frequent and bone-breaking were the falls, said Webb. By 1880 the pattern had emerged: America west of the Mississippi had two-thirds the land area of the country but only one-third the cattle. In this fabled cattle kingdom, the margins of profit and the recurrence of good years marched in step with the average annual precipitation.

Now row-crop agriculture has replaced ranching in most of the wetter grasslands of the Great Plains. Traditional ranching on the arid acres that remain produces an even smaller proportion of the nation's beef than Webb estimated in the 1930s: less than 5 percent of the national total, say some who have done the calculations.

Jerry Holechek followed the still-bucking bronc of ranching profitability from the 1930s depression onward. As in the years before, the only thing that remained constant was the inconstancy. In the depression years, "agricultural land and commodity values were much more severely depressed than assets associated with manufacturing and service." It was bad times for ranchers. Good news followed World War II, and cattle prices climbed, reaching a high in 1951. But then bad news came again. Cattle prices in real dollars (adjusted for inflation) have not been as good since.

To compound the problem of declining beef prices as the twentieth century waned, ranching costs on both public and private lands continued to rise. The most onerous costs, said Holechek, required "heavy capital outlays": water systems, fencing, changes in grazing management, brush control, the purchase of more land. In comparison, the usually slight increases in productivity resulting from new grazing schemes haven't been very important, and neither have changes in public lands grazing fees.

Holechek concluded what most ranchers presumably already knew: Profit margins of most operations have been low and are getting lower. Risks are high, and have been for many years. To survive, ranching ventures in semiarid and arid lands must be subsidized even more than they have been. Possible sources of outside money include dude ranching, hunting, ecotourism, conservation easements, and employment in town. Holding a second job now seems the mark of a rancher's loyalty to tradition.

One threat presently looms bigger than a two-hundred-pound bullrider with a jaw of chew and a black Stetson: subdivisions. The big ranches that 120 years ago were bought by wealthy people from the outside are now in danger of being consumed by tiny ranchettes, likewise bought with money from the outside.

The aging rancher fidgets. "The kids have gone, Martha," he says. "They won't be back. Jim down the road has sold out. Tell me again, what's the price that fellow offered us the other day? Bring me that telephone, and his number. Hope he's not a developer—or one of those environmentalists."

• • • • •

What drives the poor to stay on the western range, and the rich to come there? No one who has looked at the ledger-book numbers can reasonably claim the lure is economic. Walter Prescott Webb, a champion of the American cowboy, gave a perspective seventy-five years ago that still resonates:

> If the West produced comparatively so few cattle, then why is it that we think of the West . . . as the center of the cattle industry? . . . The thing that has identified the West in the popular mind with cattle is branding pen with bawling calves and the smell of burned hair and flesh on the wind! Men in boots and big hats, with the accompaniment of jingling spurs and frisky horses. Camp cook and horse wrangler! Profanity and huge appetites! . . . The East did a large business on a small scale; the West did a small business magnificently.

Only as cowboys on the western range have we been allowed so completely to reclaim our genetic heritage—our natural attraction to big animals, sweeping vistas, tribal rituals, primal odors. Here, history has given us a way to peer beneath the veneer of civilization and find out who we were so long, long ago. Other ways can take us there, but this one presently rides highest among American mythologies. The others will take time.

SEVENTEEN Resurrection

Figure 17. Saint Francis in the grasslands. Drawing by Casey Landrum.

As we pulled up to the ranch entryway, a steel barrier blocked the way: two hinged bars that spanned the cattle guard a couple of feet off the ground met in the middle. Our pickups—mine and the Bureau of Land Management truck ahead—idled for an uneasy thirty seconds or so. Then the BLM biologist stepped out, walked onto the cattle guard, and rattled the chain holding the ends of the bars together. Locked. The weather on this cloudy January day in 1995 began to look ominous.

"I'll have to radio in and have our office call the rancher to get the combination," he said. "It's a BLM access road, and I didn't expect it to be locked. Once we get through, we'll need to go by and talk to him. He's our permittee. He's got the grazing allotment."

The return call came shortly, and, after unlocking the barrier, we moved on. Why had the rancher locked the entrance? I'd thought the BLM people had advised him we were coming.

As we wound down the two-rut road, I mulled over the events that had brought me here. Several weeks before, I'd received a call from a longtime colleague of mine who had recently retired from the Arizona Game and Fish Department. Tom Waddell had taken on a post-retirement job managing a large New Mexico ranch bought by wealthy media executive Ted Turner. Tom's lifelong familiarity with livestock and ranching was coming in handy.

"Hey," he had said, "my new boss wants to resurrect prairie dogs. They used to be on this ranch but got poisoned out. Are you still in the contracting business? How much would it take for you to start us a few colonies? Send me a proposal."

After numerous phone calls and much reading about prairie dogs, I sent him a proposal. He had me come to the ranch, and we drove out to a place on the valley floor where prairie dogs used to live. He had learned about those dogtowns from an old-timer, a man who had seen the last live colony some thirty years before. On the surface of the ground we noticed pits a foot or two in diameter, which we later learned were collapsed entrances to old burrows. I learned more about the new owner's vision: "We want to put it back like it used to be."

Now here I was on the way to the largest colony of black-tailed prairie dogs left in southwestern New Mexico. Most colonies had been exterminated, I'd heard from several sources, and the ones left still attracted "varmint shooters." The species had escaped regulation by any agency in New Mexico except those controlling pests. Being on public land and not far from the ranch we wanted to repopulate, this particular colony seemed the most logical source of animals.

An Arizona Game and Fish biologist who knew about prairie dogs rode beside me. He had taken a few days' leave to advise me on trapping and handling. Now he looked around skeptically as the brush got thicker and lava rocks began to appear. "Doesn't look much like prairie dog habitat to me," he said.

Soon we sighted the ranch headquarters and slowed down. The massive mesquite-and-juniper fence that directed us toward a cluster of weathered gray outbuildings told of backbreaking labor in the desert sun. A rancher trying to make a living in this land couldn't be too pleased at having to stop work and deal with someone who wanted prairie dogs.

The rear view of the BLM biologist in front of us heightened my anxiety. His collar hid beneath his hair, and his head bobbed to a beat he must have been hearing on the earphones he wore. Beyond him, I could see the rancher waiting in the gateway—feet apart, hands on hips, blocky. Hoss Cartwright facing down a bunch of outsiders up to no good on the Ponderosa Ranch.

He proved civil but skeptical. "Why would anyone want to restore an animal we spent so many years getting rid of? How long will you be here? There's another road in to the prairie dogs; why don't you use it next time."

I breathed easier. We drove on to where the terrain opened and the prairie dog colony covered the valley floor. Thousands of mounds, many peaked like tiny volcanoes, pimpled the ground near and far. Patches of leafless mesquite and close-trimmed grasses—tobosa, burrograss, and alkali sacaton—gave promise of green to come in spring, and a lot of bare ground stretched in between the patches.

It took a few months for me to get past the rancher's suspicions that we were after something more than prairie dogs in the valley there among the lava rocks. "What are you fellows up to?" He asked the question more than once. "You and the 'Feds,'" as he called the two biologists who came with me that first day. I protested: "Only one is a federal employee, and he works for the BLM office that oversees your lease." "They're all Feds to me," he replied.

I pondered his unease. Fear of black-footed ferrets, wolves, or other endangered species came to mind. Once he made reference to "moles" among the agency people, hiding their real agendas beneath contrived ones.

Later, someone told me the Department of Defense had taken over some of his family's land when the White Sands Missile Range took shape in the 1940s. Indeed, the world's first atomic bomb had detonated twenty-five miles away, near an abandoned ranch house bearing his surname. I never inquired further. Eventually we got on better terms, the rancher and I, and by the time I left he seemed to have decided I was not some kind of mole.

During those first few months, I learned a lot about prairie dogs. A new animal noticed or a new habitat encountered seems to command a kind of intuitive perception that operates much faster than plodding science. By the end of summer I had transferred more than fifty prairie dogs to the ranch across the mountain to the west, into chicken-wire "pastures" that would contain them temporarily.

Near the end of September I pulled out of the colony for the last time, setting off the now-familiar bark and scurry of prairie dogs. I had come to like them. The landscape outside the colony, which before this summer I'd have enjoyed as a normally quiet desert scene, now seemed strangely vacant, lifeless except for the occasional jackrabbit and cow. The prairie dogs I moved to the other side of the mountain prospered and multiplied, adding new life.

• • • • •

The prairie dog project carried me into a different association with wild animals. In my youth I had followed animal sign, watched animals, shot them, eaten them, kept them as pets. For many years afterward, I had negotiated compromise between their habitat needs and the wants of humans. Now I had planted my first batch of animal seeds. It felt more like starting a refugee camp than gardening.

The work had a new name: ecological restoration. But the scientific basis for the work seemed not much different from that of producing game for hunters, assessing impacts of developers, or planning green belts for designer cities. Only the objectives changed.

Like the new label, and despite the same science basis, this work somehow felt different. Most among the biologists claiming to be restorationists assured me they indeed had embarked on a novel quest. Some of them called their work conservation biology. I heard skeptics call it playing God.

In any case, my new job restoring prairie dogs demanded that I keep current on restoration theory and practice. First of all, I needed to understand the rationale for what I was doing. But the cloud of new information dazed me, not only by its bulk but also sometimes by its haziness. Especially disconcerting to me, some restorationists seemed to call upon the same belief in a Grand Design that a century earlier had given rise to succession theory. To help sort things out for myself, I began to classify restoration ideas as "faith-based" or "science-based."

Those espousing faith-based ideas seemed to be asking me to trust them. They claimed allegiance to science but then went beyond logic to create a kind of evolutionary god's vision. They used words I had a hard time understanding: *natural succession, integrity, development trajectory,* and so on. They preached good and bad, degradation and recovery, shoulds and should nots. Some of the more insistent reminded me of that long-ago preacher who claimed inside knowledge of God's will.

My favorite science-based planners claimed no all-knowing Umpire to sanctify their visions. Instead they called upon reason and admitted bias. Some said, "We think rare species and wildlife diversity need conserving to preserve options for people yet to come." I thought, "Prairie dogs make rangeland more interesting than it was to me and other kids." It felt better, knowing the viewpoints and understanding the words.

In 1982, the University of Wisconsin started publishing a small journal called *Restoration and Management Notes.* Now called *Ecological Restoration* and matched by a companion journal, *Restoration Ecology,* this journal and its founding institution gave rise in 1988 to the Society for Ecological Restoration. In 2004, the society published its second edition of a primer on ecological restoration.

The restoration primer called for bringing back biological diversity. It recommended using historic baselines as templates. These ideas sounded reasonable, or at least in line with my biases, so I listened.

Then the primer spoke of ecosystem health and integrity, and the need to measure them. It warned against ecological degradation and similar sins. I got lost in a thicket of new terms, a jungle of confusing metaphors. The language befuddled me almost as much as the marketing of succession and climax had in my university years. It contained a new version of a Purposeful Nature, of an ecological system that could get back on the right track with a little help from scientists who had the inside scoop on right and wrong.

I hesitated. By now I knew that one animal's heaven is another's hell. Climax grass looks good to cows but not necessarily to goats, deer, quail, or prairie dogs. Which animals did these restorationists favor, and why? The society's primer might indeed have merit, but I needed clearer language and less of a "trust us" tone to its recommendations.

My father's skepticism emerged again from the distant past. "Be careful who you believe." By this time I'd decided that, in the growing field of "environmentalism," the charlatans and the self-deluded were multiplying faster than the careful scientists. I wanted to understand the words, see the evidence, and hear the biases.

• • • • •

Soon after my first adventure seeding prairie dogs, I found myself in the business of restoring more of them. To my good fortune, a dedicated and well-informed cadre of biologists had already blazed the way. Gradually I became accepted by the tribe, indoctrinated into its rituals, and intrigued by its habitat. Like me, these folks sought to resurrect a lost Eden.

But my search for how it used to be—the grassland template with prairie dogs embedded—turned out to be more complicated than I'd at first imagined. Initially the only thing that mattered was, Were prairie dogs here before? A "yes" answer could come from several sources—memories of old-timers, records of poisoning campaigns, visible signs of long-abandoned dogtowns—and meant we could and should put them back. A "no" answer meant forget it; prairie dogs don't belong.

But then a disconcerting question arose: *When* before? Like many of my colleagues who lamented the passing of the American frontier, I'd vaguely assumed the ideal ecosystem state—the baseline for restoration—had existed at the time of the first "coming of the white man." Others working toward conservation of wildlife often seemed to agree. That assumption worked fine until I began to look for details in the template.

Two circumstances coalesced into a problem. First, for most regions in or near historic prairie dog range, I had trouble dating the white man's "first coming." Second, and more perplexing, the distribution of prairie dogs seemed to have been in great flux over time. Needing grazing to prosper, they had diminished greatly with the slaughter of bison. Then, with the coming of cattle, their numbers had exploded. Unfortunately, most records delineating their historic range came from the period of flux and, to make matters more confusing, from different times in the period. Faced with this problem, I began to select reintroduction sites more on the basis of present-day grasses and grazing than on conclusive evidence that prairie dogs once had lived there.

The years passed. I saw black-tailed prairie dog colonies from Mexico to Canada change and be changed by their habitat. Simple answers didn't work; ugly uncertainties continued to arise. Why didn't prairie dogs introduced to sites of extinct dogtowns always "take"? Why did colonies devastate the grass in some areas but less so in others? How did the setbacks in "succession" generated by prairie dogs fit with the long-assumed equivalency of "pristine" and "climax"?

Printed information proved often inadequate and internally conflicting. Ecological findings from one area extrapolated poorly to others. Historical accounts misled. Evidence from the archeological and paleontological sciences raised new questions. Out in the field—that ultimate testing ground—the prairie dogs themselves often contradicted written accounts and restoration mantras. But some patterns did turn up.

Prairie dogs need short grass. Only if they can see over the grass do they cope well with the deadly impact of coyotes, bobcats, and other stalking predators. They don't particularly care what keeps the grass short—cattle grazing, fire, or even a tractor pulling a bush-hog work just about as well as grazing by bison, the historic prairie dog facilitator. And they don't care whether the short grass is within their historic range or not. I learned that introduced colonies had thrived as far away as Nantucket Island off Cape Cod.

Prairie dogs reduce rangeland "condition." They graze very intensively. As colonies age, the grass cover within colonies declines and the condition trends from good to poor on the traditional rangeland manager's scale. The vegetation inside old colonies never ranks anywhere near climax.

Prairie dogs benefit rare wildlife. The black-footed ferret, often billed as North America's most endangered mammal, needs dogtowns to survive. Burrowing owls, ferruginous hawks, mountain plovers, and swift foxes fare better inside than outside colonies. The list goes on. Why the singular attraction of dogtowns to rare species? Most likely because the unique combination of short grass, disturbed soil, deep burrows, and large poundage of prey in small chunks appears nowhere else in grasslands managed for so long by the rules of Frederic Clements.

• • • • •

My final conversion to the science plan came when the god of coevolution died. The paroxysms I noticed that eventually would prove fatal seemed invariably to arise in the company of those who called themselves paleoecologists, students of long-ago associations between animals and habitats. They seemed caught up in a time warp.

My 2004 edition of the *Oxford Dictionary of Ecology* defines coevolution as "a complementary evolution of closely associated species." Its central premise—that long evolutionary associations create hard-wired dependencies—forms part of the rationale some invoke for restoring whole ecosystems in preference to a kind of parts supply house that makes repairs one species at a time. The implication seems straightforward: species that evolved together belong together. In my earlier years it made sense, and on this basis I applauded the "ecosystem restoration" idea.

But then the paleoecologists pointed out the flaw in the logic. Most species are far older that the communities they now occupy. This implies a constant shifting of species associations over time. The evidence comes from layers of earth laid down through the millennia, from fragments of ancient plants and animals in dry caves and packrat middens, and from pollen in sediments of ancient lakebeds.

This left me dumbfounded. If indeed species associations had continually changed, coevolution belonged back on the shelf of interesting but irrelevant ideas.

I hoped it wasn't true. Reassembling a community's lost fauna and flora offered excitement akin to resurrecting Pharaoh's pyramid from scattered stones. I looked for evidence to rekindle my belief. Aha! Black-footed ferrets in prairie dog towns. Everyone agrees they need to be there. Doesn't that imply coevolution? Probably yes, the paleo-people admitted; but that kind of close-coupled arrangement turns out to be more the exception than the rule.

The case began to close for me in 1995. That was the year I first met prairie dogs up close and the Society for Range Management buried Clements. The final persuasion came under a ponderous title: "Pliocene-Pleistocene Biogeographic History of Prairie Dogs, Genus *Cynomys* (Sciuridae)." The article appeared in the *Journal of Mammalogy* in 1995, though I didn't read it until a few years later. It had been authored by one H. Thomas Goodwin.

The article made clear Goodwin's obsession with time past. He'd pored over a stupendous number of museum specimens and records of bone fragments and teeth. He'd compared ancient samples with modern ones, thus tracking the origins and distributions of different species of prairie dog. He'd plotted collection locations of prairie dog remains on maps and by date.

The short of Goodwin's story looked like this: Black-tailed prairie dogs have lived in the central grasslands of North America since long before the last great continental glaciation—the Wisconsin—receded ten to fifteen thousand years ago. Their anatomy, and by inference their behavior, have not changed. Their remains thus confirm the presence of grassy habitats in those faraway times. These clues, and other studies by paleobotanists and paleoclimatologists, show the Great Plains to have stretched more or less between the Mississippi River and the Rocky Mountains for hundreds of thousands of years. Spruce, fir, and other trees sometimes intermingled with grass, perhaps in a mosaic or savanna pattern.

Something looked very odd about one of Goodwin's maps. In the late Pleistocene, black-tailed prairie dogs ranged farther eastward than they ever did historically—into central Iowa, and to the southeastern coast of Texas. But the climatologists said the Great Plains were wetter then than now. A wetter climate should have meant taller grass, especially in the east, where the shape of the continent would even then have brought more rain. Prairie dogs hate tall grass. So how could prairie dogs in the Pleistocene have lived in what almost certainly was tallgrass prairie?

A tempting answer looms in the great diversity and abundance of large grazers back then. Those who have unearthed the ancient bones from sinkhole and floodplain say more than a dozen species larger than a modern bison or cow roamed the prairies at the time. Now-extinct horses, camels, long-horned bison, ground sloths, mammoths, and others grazed, trampled, and wallowed, making short grass of tall grass and dust bowls of fragile soils. It was prairie dog heaven.

The American bison, that leftover dwarf of the Pleistocene, must exert a pitiful disturbance indeed compared to what happened before. Ten thousand years represents an evolutionary instant, not discernible in the prairie dog bones Goodwin sorted in dry museum trays. If my prairie dogs yearned deep down for anything more than bush-hogs or bison, perhaps it was for greater beasts, more trampling, shorter grass, and dust storms raging on across a tattered plain.

• • • • •

Sociologists say that every human needs a cosmology, a belief system to give purpose to his or her life. Folk myths, group religions, family values, and tribal patriotism all offer belief systems that save a secure place for the individual. The philosopher Eric Hoffer noted the consequences of discarded belief systems: something else will be adopted to fill the vacuum. The Beat Generation, the Jim Joneses, and the perpetrators of the Holocaust all gained recruits from among the disenchanted. Some had better things to offer than others.

The problem with applying logic and reason to belief systems is the uncertainty that arises when beliefs prove unfounded. Over the years, deaths visited on my small gods by science have summoned periods of mourning. Lost beliefs have generated episodes of aimless uncertainty. Like a person bereft of partner, it has been tempting to pick up the first substitute that came along. It is not easy, being patient.

EIGHTEEN Pleistocene Park

Figure 18. Rancho La Brea diorama. Courtesy of the University of California Museum of Paleontology, Berkeley.

On August 18, 2005, the prestigious journal *Nature* published a grasslands restoration vision so bizarre that some took it as a joke. Featuring a dozen authors, mainly from universities across the United States, it aimed at people who already had heard so many novel ideas for conservation as to be severely desensitized. But the *Nature* article startled them awake.

Responses began pouring in even before the paper issue made it to subscribers' mailboxes. Some were complimentary, some angry, many incredulous. Within a few days after publication, emails and phone calls to the main author had climbed into the hundreds. Commentary appeared in most of the mainline print and television news outlets in the United States, in the Associated Press, and in major papers on all continents. The concept put forward appeared among the *New York Times Magazine*'s "Big Ideas of 2005."

Few scientific papers can trace their origins to very specific times and places, but this one could. The time: September 10–12, 2004. The place: Ladder Ranch Lodge in the eastern foothills of the Black Range, southwestern New Mexico. I sat in the meeting room of the lodge and listened to the authors wrestle for two days with the idea that hit the pages of *Nature* eleven months later.

The front porch of the Ladder Ranch Lodge faces eastward, and the view draws your eyes down Animas Creek Canyon. Cottonwoods line the creek bottom, abrupt cliffs rise on the south side of the canyon, and prickly pear cactus and mesquite trees climb the less precipitous slopes on the north side. Animas Creek meets the Rio Grande ten miles downstream.

Late in the afternoon of September 10, I could see an occasional dust plume ascending the canyon toward the lodge. Each boiled up from the east and approached on the road that paralleled the creek, settling only as a vehicle pulled up to the porch. By dark most had arrived: Paul Martin from the University of Arizona, Harry Greene and Josh Donlan from Cornell, David Burney of Fordham University, Joel Berger of the Wildlife Conservation Society, and a half dozen others.

Most of the visitors already knew, or soon noticed, that the Ladder Ranch ran bison, not cattle. Elk and mule deer fed in bottomland fields in view of the lodge. Each evening javelina shuffled out of the mesquite thickets fringing the creek, and manager Steve Dobrott spoke of problems with cougars hanging around headquarters. The ranch offered a native cast of restoration characters seldom matched from one vantage point in the American Southwest. But the scientists gathered here had their collective eyes on bigger quarry.

"Re-wilding North America" ran the title in *Nature*. The byline that followed hinted at the unthinkable: "A plan to restore animals that disappeared thirteen thousand years ago from Pleistocene North America offers an alternative conservation strategy for the twenty-first century, argues Josh Donlan and colleagues."

Beyond the hook, the rhetoric became more rational. A host of large North American mammals—herbivores and their predators—abruptly had gone extinct at the end of the Pleistocene. Overhunting by humans seemed at least partly to blame. Large mammals in Africa and Asia, some very similar genetically to those that disappeared here, face a bleak future as human impacts burgeon. Why not replace some of the extinct North American species with related Old World forms—ecological surrogates as it were—in need of protection?

What species are we talking about? Well, take horses and camels, for example. True New World natives, these groups evolved here millions of years ago. Later they spread to Asia and Africa across a Bering Land Bridge that then connected the Old and New Worlds across what is now the Bering Sea. At the close of the Pleistocene they abruptly disappeared from North America, the continent of their birth. Why not bring back to North America, in a kind of repatriation, some of the present-day Eurasian forms—Przewalski's horses and Bactrian camels, for example—that are in danger of extinction?

What about elephants, as surrogates for the mammoths that disappeared ten thousand years ago? They might be useful to help combat the brush that recently has invaded so many western grasslands. In Africa, elephants push back or thin vast woodlands to the benefit of grasses, and probably always have done so. Recent studies show that the Asian species of elephant more closely resembles extinct North American mammoths than it does the African elephant. Maybe we could start with a family group of semidomestic Asian elephants managed by an Indian mahout.

And then there are the big predators. Now-extinct cheetahs that roamed the Great Plains of the Pleistocene resembled closely the ones that live in Africa today, and most experts consider the American Pleistocene lion the same species as today's African lion. Once we restore large grazers, doesn't it make sense to bring in predators that can control them? Maybe we'd need to prepare the public in advance.

Looking back, I recall that the predator proposal stirred the most debate at the workshop. I still can see Dave Foreman, founder of Earth First! and apparently not much awed by the learned scientists around him, waving his hands and shouting, "Imagine cheetahs chasing pronghorns out across the Wyoming sagebrush!" "Wait," said the more pragmatic. "Public fear of big predators could torpedo the whole idea."

The media frenzy that followed the original publication didn't help. Reporters knew they could sell copy with images of lions in Nebraska. "Crazy environmentalists," muttered some readers.

The Re-wilding Dozen had come to the Ladder Ranch with more academic zeal than practical experience in translocating and managing animals. They did mull over how it eventually could be done. People and the conventional ranch lifeway are exiting the western Great Plains, some said. What a place for a Pleistocene Park! Maybe we could start on a large private holding. Public lands could come later. How do we get around the antipathy to anything called exotic? How do we deal with the tendency for rural people to be conservative? But let's worry about these things later—right now, let's focus on making the idea reasonable.

• • • • •

Most men attracted by large eyes and big breasts may never have heard the term *superstimulus,* but their genes know what it is. Forty years ago at university, I learned of Konrad Lorenz's early experiments that showed the tendency for brooding birds to select oversized eggs in preference to their own. A superstimulus is an exaggeration of an image, taste, sound, or odor important to a species over evolutionary time. It causes a heightened response. Product manufacturers capitalize on it when they lace junk foods with sugar and fat and accentuate speed and size in automobiles.

On the human evolutionary stage, animal size mattered a lot. The larger the size, the greater the human response, and supersized counted most. The biggest predators were the scariest. The largest grazer carcasses served best to feed the clan and confer prestige on the hunter or scavenger who provided it. Dead animals have since gone out of style as female attractants, but the men haven't realized it yet and a faithful cadre still advertises male prowess with large stuffed heads on living room walls. Great predators no longer threaten the gene pool, but movies like *Jaws* and *Jurassic Park* still call forth the superstimulus response and the box-office dollars.

Zoos and wildlife parks seeking public support rely on their charismatic megafauna. The term seems almost redundant. Large animals nearly always exude charisma; they attract a large proportion of the human interest and bring in most of the dollars for conservation. Few people I know go on nature tours to see rodents or weasels. Some do go to the far corners of the earth to watch birds, but in that case other stimuli important to our early survival—fascination with color, joust for status—may be at work.

On a more pragmatic level, we need the digestive abilities of large herbivores to give us a comfortable living in rangelands. This habitat covers more than half the earth's land surface, and that coverage will only enlarge as we run increasingly short of the fossil fuels and fossil water that have let us farm vast areas otherwise too dry. The question is not whether we will always need or want large animals. Rather, it is: Which animals?

This question will find a better answer in the field of human preference than in animal ecology. The animals can adjust. Historical experiments in moving horses, donkeys, cattle, sheep, water buffalo, American bison, and others around the globe demonstrate what nutritionists announce with confidence: grazers bigger than a golf cart seldom have met a range they didn't like.

The authors of the *Nature* article spoke of seeking "surrogates" for lost species. They proposed matching mammoths with elephants, extinct camels with living ones, and so on. As demonstrated by inadvertent and controlled experiments worldwide, ecological surrogates might take much different shapes and lineages than the authors had in mind. Here again, the emphasis on replacement with close relatives may have more to do with what people will accept than with what the megafauna will.

This crystallizes the real problem with Pleistocene or any other kind of re-wilding–something that history illustrates well. It is the fickle tastes of that most influential of the large grassland species: us.

• • • • •

Projects in "re-wilding" have been going on for a long time in Texas. The first may have been that of Captain Alonzo de León, who in 1690 intentionally released European cattle at likely spots along his expedition's travel route from Mexico to the missions around Nacogdoches in what is now eastern Texas. These seedings expanded; more came with later expeditions from Mexico; wolves and other predators culled the unfit; and by the early 1800s central and southern Texas sustained a hardy wild animal of European descent—the longhorn cow.

The Anglo-Americans who took political control of Texas in the 1830s soon pushed cattle into vast new areas. Hardy drovers such as Jesse Chisolm and Charles Goodnight took great phalanxes of them west, east, and north. The so-called Texas longhorn went on to populate much of the western grasslands, and the rest is dime-novel history.

In Texas itself, longhorns were only the beginning of re-wilding. In the 1920s, the great King Ranch in South Texas started a move that became a Texas passion: running exotics. By then, of course, cattle stayed mostly behind fences and few thought of them as exotics, just livestock.

The King Ranch example started a stampede. By 1963, ranches in the state carried an estimated 14,000 wild nonnative hoofed animals of thirteen species other than cattle. By 1994, the Texas Parks and Wildlife Department estimated the total number of free-ranging exotics at 77,000, nearly all in the temperate to subtropical savanna country of Central and South Texas. As the century drew to a close, Texas had several times the number of exotic mammals as all other states combined.

By this time the number of exotic species in the state had increased to an astounding seventy-one. Most remained rare and secure behind ranch fences. Elizabeth Cary Mungall, known for studies of blackbuck antelope and other exotics in Texas, assembled census figures. Among the more abundant and widespread were axis deer, 55,000; Indian blackbuck antelope, 35,000; fallow deer, 27,000; Barbary sheep, 12,000; sika deer, 12,000; and nilgai antelope, 8,000. David Schmidly, a well-known Texas mammalogist, considered at least these six sufficiently numerous and uncontrollable to be permanent additions to the state fauna. They and others continue to expand.

In all fairness, the Ladder Ranch re-wilders had mostly bigger things in mind. Paul Martin saw five-ton elephants as the ultimate goal, though most at the workshop seemed to think horses and camels might be more manageable at first. In Texas, the largest of the presently abundant species—the nilgai antelope—tops the scales at around 600 pounds. But bigger ones may be on the way. Elizabeth Mungall's list for Texas ranches shows several hundred zebras (in the half-ton range), a hundred giraffes (approaching two tons each), and fifty or so rhinoceroses (one to three tons each, depending on species).

So what's the difference between Pleistocene re-wilding and Texas exotic ranching? Ecologically, maybe not much. Indeed, the re-wilding folks recommended a closer look at what's happening in Texas to inform the way for wider applications of their ideas. But in the public psyche, there may be a lot of difference. Some see re-wilding as an option for how people around the world can rethink their relations with other species. Some of the same group dismiss exotics ranching as just a bunch of rich Texans playing with animal toys.

How people respond to the Texas experience may suggest whether the re-wilding idea is nutty or what. Elizabeth Mungall touches on some practical objections: risks to native species (disease, competition, interbreeding), range degradation, costs of management. And then there are the ethical objections: interference with the natural order of things, homogenization of world ecosystems, and messing up the native American template. Ranchers running exotics, however, note the potential monetary profits, the good that can be done for conservation of rare species, and the pleasure that can be had.

Texans have enjoyed some pragmatic advantages that may not play out elsewhere. They have the habitat—the warm and productive savannas capable of hosting numerous herbivores on the same landscape. They have an exaggerated private ownership ethic that turns large ranches into small kingdoms impervious to public or state control. Not least, they have seen a profitable century of pumping oil and so have the cash, at least for now.

. . . . .

If you take Interstate 25 north from Las Vegas, New Mexico, or south from Raton, you will not travel far before you see a peculiar monolith rising up on the horizon ahead. Contrasting with the shortgrass prairie that rolls mile after mile to the right and left, it resembles the side view of a prairie schooner. American frontiersmen on the Santa Fe Trail from St. Louis to Santa Fe called it Wagon Mound. They camped at its base, where the present village of Wagon Mound offers a small choice of food and gas.

On a green prairie day near the very end of the twentieth century, Paul Martin and I drove east from Wagon Mound in search of the Gruenwald Ranch. Paul would later play a key role in the Ladder Ranch re-wilding workshop. I had known him for several years, and of him for many more—as originator of the "Pleistocene overkill" hypothesis to explain the demise of North America's megafauna ten to twelve thousand years ago. We were on the way to see Mr. Gruenwald's horses.

Mark Jansen, the veterinarian in charge, met us at the ranch headquarters. He showed us the office with it horse paintings and sculptures, and the attached hangar the owner used when he flew down from Colorado Springs. We talked of the original breeding stock—mostly excess animals from Germany's Stuttgart Zoo, horses of various species that Mr. Gruenwald began acquiring in the 1970s. The motivation? Maintenance of a gene pool of rare equids for conservation options. The future? Uncertain; no heirs were waiting to carry the project forward.

Then we drove around the ranch to see the most unusual assemblage of free-roaming grazers on blue grama I ever expect to see. Przewalski's horses with dense upright manes, Iranian donkeys swarming the shortgrass like rabbits, Grevy's and Burchell's zebras from Africa, and a few less common species in corrals. "Oh, yes, they all thrive on the native grass," said Jansen. "We supplement those out on the range with hay, now that they've multiplied beyond the carrying capacity of our six thousand acres. No, we haven't found many people anxious to buy the excess. Yes, it would be good to get some research going."

Invasion of the aliens, or return of the natives? "It all depends on your time perspective," said Paul. "Living as I do in a Pleistocene world, they seem right at home to me."

Feral versions of the ordinary domestic horse have roamed free in the Americas since shortly after the first ones escaped the caravans of Cortez and Coronado. The Great Plains tribes gloried in them. Later on, the cattlemen complained about them and the rest of America waffled. We seem for now to have settled on the unstable notion that a few wild horses are okay out there on the western range, but we don't want too many.

. . . . .

In September 2005, a small group of New York City people traveled out to meet with another handful of people several miles west of Watrous, New Mexico. Unlike stereotypical big-city folks, most came wearing blue jeans and rumpled shirts and carried small duffel bags rather than Samsonite cases. They gathered at a ranch called the Wind River. They carried a grasslands restoration idea much more impressive than their visible baggage: they wanted to bring back the American bison.

What a sensible idea, some might say. Forget the surrogates from Africa and India. Why not focus on the biggest Pleistocene animal to make it through the great extinction bottleneck. Re-wild with bison.

But what was a bunch of easterners doing out on the New Mexico prairie, meddling again into western affairs? Why should such a notion have emerged so far from the great central grasslands where Native Americans chased bison more or less since the Pleistocene, and European Americans almost exterminated them a century ago? Why New York? The answer lay with a momentous anniversary and conservation's most important president. Exactly a hundred years earlier, in 1905, Theodore Roosevelt had gathered together a few cronies at the Bronx Zoo in New York City and set up the American Bison Society. The purpose: to save an American icon in risk of extinction. The number of bison had plummeted from perhaps thirty million in Lewis and Clark's time to a mere several hundred at the beginning of the twentieth century.

The folks who assembled a hundred years later at the Wind River Ranch represented the Wildlife Conservation Society. The WCS operates as part of the Bronx Zoo. Affiliated with the American Bison Society from the beginning, it became the heir-apparent bison protector after that society disbanded in 1950.

Was the gathering at Wind River any more than a self-congratulatory back-thumping for a species saved? After all, half a million bison now grazed American grass. The species seemed out of danger from the threat of extinction.

But as the meeting got under way, those in charge pointed to a second and perhaps greater threat. The American bison seemed to be following the European aurochs to domestication. Ninety-five percent of those alive could count on feeding domesticated people. Most lived their grazing lives out in small pastures, and thereafter would be pushed through cramped feed lots to be "finished" on rations designed originally for cattle. Genetic signatures showed that past interbreeding with domestic cattle had contaminated the DNA of the great majority of bison now alive. Most alarming, some breeders selecting for "desirable" conformation already were producing bison shaped like cattle.

What to do? Consensus among the Wind River Ranch guests denounced close-corralled domestication. A Wind River backdrop of juniper savanna and canyon, coupled with the sweeping view westward toward the high Sangre de Cristo Mountains, pulled a wilderness vision from most.

The following year, the Wildlife Conservation Society helped assemble a larger group of bison enthusiasts at the Vermejo Park Ranch near Raton, New Mexico. Rhetoric became more strident. Determination built. A sweeping vision emerged from the WCS people: "Over the next century, the ecological recovery of the North American bison will occur when multiple large herds move freely across extensive landscapes within all major habitats of their historic range, interacting in ecologically significant ways with the fullest possible set of other native species, and inspiring, sustaining, and connecting human cultures."

A lot can happen over the next hundred years to threaten this vision. Human populations may double. Oil scarcities almost certainly will cripple livestock production operations on many ranges. Stable governments may topple. Economies will collapse. How will the folks out in the arid grasslands continue to feed their bodies and minds?

Large herds moving freely? Inspiring human cultures? One can hope. All the evidence suggests we will be in need of inspiration. Experiments in Pleistocene re-wilding may take turns we never could have imagined.

NINETEEN Diversity

Figure 19. Rangeland biodiversity. Drawing by Casey Landrum.

Encountering diversity in nature can titillate the senses but at the same time generate anxiety. It's like owning a lot of stuff or having too many people or pets in the house. Diversity offers grand prospect, interesting opportunity, but at the same time can challenge one's sense of control. Birdwatchers spend a lot of time seeking out diversity, lawnkeepers and farmers a lot of money getting rid of it.

Conserving biodiversity justifies a lot of political rhetoric and management action these days. What is biodiversity? Biologists squabble endlessly over definitions and perspectives, but getting too detailed can be confusing. So we'll stick with the two dimensions important to most people: number of species and habitat variety. Ecologists sometimes refer to these as species composition and habitat structure. Whether a given acre has one, ten, or a hundred species, and whether it is cropland, grassland, forest, or a mixture of types, means a great deal to humans and other inhabitants.

Why do conservationists so bewail biodiversity loss? Partly because it's in vogue to do so, but more significantly because our species depends on diversity in nature. In the narrow view of daily life, it may seem we don't need more than a few plants and animals and can get along without a lot of landscape variety, but a closer look shows a phenomenal array of species and habitats we used in the past, a surprising variety we use now, and a widespread unease with losing more than we've already lost. Without doubt, unforeseen problems will arise in the future. We always will need a pool of diversity from which to seek solutions. To purposely exterminate any organism less offensive than a human tapeworm seems akin to playing God, and a pretty shortsighted god at that.

Does the modern human lifeway pose an unprecedented threat to biodiversity? Yes, says Timothy Swanson, economist at Cambridge University. You'll find the danger embedded in industrial capitalism.

In 1995, Swanson assembled a group of technical papers on the confrontation between biodiversity and economic development. Cambridge University Press published these in a book, *The Economics and Ecology of Biodiversity Decline.* In one of the articles, Swanson himself explained why development threatens biodiversity conservation.

Production efficiency, Swanson says, improves with the uniformity of the resource being developed. For example, growing a single crop such as corn or soybeans yields a lot more profit than does producing several different crops in the same area. This is true because the entire infrastructure—tractors, fertilizers, pesticides, herbicides, transportation fleets, processing facilities, marketing strategies, and so on—can focus on one particular commodity.

Expanding this concept in time and space means that uniformity in food production is quickly expanding around the world and into the future. And as competing crops and competing methods of cropping fall by the wayside, the diversities of agricultural machines, handling procedures, and food plants shrink in unison.

We already see some of the consequences. Four plant species—corn, wheat, potatoes, and rice—in a handful of high-yielding varieties now make up the majority of all human food (though you might not guess this from the infinite guises of corn and wheat on the grocery shelf). The total diets of industrialized nations today include only a tiny fraction of the ten thousand or so plant species eaten by humans over historical time. In the jargon of ecology, we have become specialist feeders.

A mere twenty years before Swanson's assessment, Philip Smith of the University of Montreal had noted increasing complexity in crops. Exchanges of crop species and varieties "between different regions and different continents in the past few millennia," he said, had elevated local crop diversities almost everywhere. Now it seems that globalization in economics and food production has reversed the trend.

Our tendency toward specialization may have ancient roots. Most aboriginal tribes I have read about used a lot of species but got most of their sustenance from only a few plants and animals. Among North American hunters and gatherers, Great Plains tribes followed the bison, eastern tribes concentrated on white-tailed deer and forest nuts, and the Apache in Arizona and New Mexico ate a lot of deer, acorns, and agave.

Once agriculture took root—from a few hundred to several thousand years ago, depending on location and tribe—those who embraced it tended to become even more specialized than hunters and gatherers. In most of the Americas the specialty became corn, and people grew squash and beans and hunted small animals as subsidiary occupations. The several other grass species originally domesticated in the New World left most fields for good once people realized the advantages of corn.

The drive toward specialization in prehistoric cultures probably came from the same inclinations that propel it today. Nature is extremely complex, and, despite the phenomenal learning ability of our species, we as individuals never can know very much about this complexity in total. So we focus on the resources that seem most useful at the time. This generates an individual knowledge, and eventually a cultural one, that enables people to get the most calories and to otherwise thrive in a particular time and place.

The farming and ranching culture that surrounded me in youth followed me into older age. Its totems of subsistence—cattle, horses, dogs, corn, and beans; and sometimes hogs, turnips, or chiles, depending on local climate and preferences—have not changed much. Fortunate enough to get paid to explore the complexities of life beyond the totems, I continue to be struck by most people's disinterest in biological diversity. "What good is it?" they commonly ask. My answer—that it may be useful someday to someone else's grandchildren—seems unimpressive in the context of the more immediate challenges of flood, drought, or joblessness. Looking too far beyond the next meal probably did little to improve the success of their lineages or mine.

My grandfather showed a surprising knowledge of the diversity of trees, given his economic focus on part-time jobs, small-time ranching, and hunting of feral hogs. I did well at university in dendrology—the study of trees—primarily because Granddaddy already had taught me the names and uses of trees in our forest. As it turned out, his knowledge probably sprang from utility. He came from a subsistence lineage that put to advantage the useful qualities of many tree species, and before my time he had managed a logging crew in the East Texas woods. But he couldn't name very many small birds, insects, or grasses.

• • • • •

Dr. Sam Fuhlendorf, range science professor at Oklahoma State University, spreads odd notions about ranching in a part of the country where unconventional ideas get the same critical scrutiny as a cow wearing an altered brand. We first met during a visit I made to Oklahoma at the end of the last millennium. Hailing from my home state and like me a product of Texas A&M, Sam looked and talked like one of the boys. But he turned out to be a maverick.

One of the most egregious sins committed by Sam came to my attention in an article he published with colleague David Engle in 2001 in the journal *BioScience.* Packed with terms and concepts seldom tossed about in cowboy culture, it challenged what livestock growers had been doing out on the range for most of a century. Fuhlendorf and Engle titled the article "Restoring Heterogeneity on Rangelands: Ecosystem Management Based on Evolutionary Grazing Patterns."

The introduction gave the theme: "We propose a paradigm that enhances heterogeneity instead of homogeneity to promote biological diversity and wildlife habitat on rangelands grazed by livestock." The ugly details followed. Traditional rangeland management, they wrote, reduces diversity by favoring the singular needs of cattle. Maximizing the flow of energy through cows, long touted as ecologically best, has deleterious effects on most other animals. That needs to change.

How, then, should range be managed? Instead of striving for uniform stands of cow food and uniform grazing fence to fence, we should encourage spatial heterogeneity. The key is to manage for variety in plant species, plant structure, and grazing pressures within pastures. Forget climax. Think patchwork. Embrace disturbance.

Won't this be a lot of trouble? Not really, said Fuhlendorf and Engle. Cows, bison, and indeed all large herbivores naturally graze patchily. Conventional managers over the years have seen how much trouble it is to get them to graze uniformly. You can enhance their natural tendency for patch grazing by introducing patchy disturbance. Burning, for instance, attracts grazing for several years afterward. Periodic patch burning sets in motion a self-perpetuating and shifting mosaic.

• • • • •

Bob Hamilton practices on the ground what Fuhlendorf preaches in the university classroom and at rangeland conferences. Bob manages the Tallgrass Prairie Preserve in northeastern Oklahoma. The preserve, owned by the Nature Conservancy, runs bison instead of cattle. It sits at the southern end of the Flint Hills, which because of the shallow, stony soils have largely escaped the plow and sustain the biggest tallgrass region left in North America.

Each year Bob burns a lot of tallgrass, maybe a quarter to a fifth of the 38,000-acre preserve. He burns different patches each year, so that on average any one site gets burned once every several years. Bob says fires swept through about as frequently when the Pawnees and Comanches hunted bison there. Burning takes out the aboveground parts of big bluestem and other tall grasses. It makes way for a temporary flourishing of broad-leafed herbs—called forbs by Bob and other range ecologists, and weeds by the old-timers. The dominant grasses grow three to six feet tall and, unless burned periodically or grazed heavily, tend to shade out most other plants. Having evolved with fire, they resprout from rootstocks within a few weeks or months after fire, depending on the season.

Bob burns mostly in the spring. The bison seek out the better-quality new growth in preference to the unburned, and soon Bob has a natural grazing rotation and moving mosaic going. And he avoids the expense of cross-fencing.

Bob grew up in the region. "It's a lot of trouble to get local people interested in our system," he says. "They're stuck with tradition. They run cattle. They supplement in winter to make up for the poor nutritional quality of the tall grasses outside the growing season. That's what they've always done, and I guess it's hard for them to change. We burn, we don't supplement, and the bison produce a lot of calves and grow fast.

"I suspect," Bob says, "that running bison and patch-burning grass produces meat more economically than conventional cattle grazing. Fire takes out the standing dead, or past years' growth, which in tallgrass regions loses quality in winter and dilutes every mouthful an animal takes. Furthermore, the metabolism of bison slows down in winter more than that of cattle. This may be an evolutionary adaptation to northern climates. Consequently the bison seem to do about as well in winter as cattle but with less and lower-quality food. The bottom line: we get by without supplemental feeding in winter, and the cattle growers don't.

"What's more," he says, "the patchwork generated by burning and selective grazing attracts species not benefited by uniform stands of tallgrass.

Deer follow the burns. Prairie chickens seek out the shortgrass for breeding and feeding and hide in the taller patches nearby. They and bobwhite quail need the forb and grasshopper production that follows burning and grazing. Different stand heights attract different species of the smaller birds and mammals. Overall, we get a more diverse mix of both plants and animals than we would with conventional grazing."

Fuhlendorf conducts range research projects on the Tallgrass Preserve. He and Hamilton collaborate in the effort to crack tradition. It's a hard nut.

Their approach builds on the accumulated wisdom of evolution. Part of that wisdom is patience. Change will not happen overnight. New approaches may not always appeal to old-timers, but ranch by ranch, old families move away and newcomers take their places. Many of these—the Nature Conservancy people among them—seek more than just another dollar from a domestic converter of energy. They are looking for a vignette of the past, for something more like the pre-cattle diversity in plants and animals.

• • • • •

A five-hundred-mile drive heading due west from the Tallgrass Preserve shows you a good cross-section of the southern Great Plains. You can take U.S. Highway 60, which eventually turns into 64. Once you leave the Flint Hills and the soils deepen, vast farms swallow up most of the landscape. Stop randomly and look out; sorghum, wheat, or cotton usually will bring the diversity of species at any given location to near one. By the time you reach the eastern boundary of New Mexico, you will have crossed, in succession, the prairies wet enough to farm without irrigation, and the Ogallala Aquifer that lets Texas and Oklahoma panhandle folks farm with irrigation where they otherwise wouldn't farm at all. Once in New Mexico, once past the wetter eastern prairies and the water subsidy from the Ogallala, native grass rolls once again to the far horizon. Diversity replaces uniformity. Drive on and, near Raton, the Great Plains end.

Thirty miles southwest of Raton looms another project in biodiversity restoration. The annual precipitation here is one-third that on the Tallgrass Preserve and the grass less than one-third as tall. Cowboys and most range managers call it shortgrass prairie; academics prefer the name shortgrass steppe. Here, through a great expanse of blue grama, winds the Vermejo

River—barely a creek by eastern Oklahoma standards—and on its cottonwood banks lives a wildlife biologist named Dustin Long.

Dustin works for the Turner Endangered Species Fund. Since about 1998 he and his family have lived here on the historic Vermejo Park Ranch, bought by media executive Ted Turner in 1996. The 570,000-acre spread reaches west to the crest of the Sangre de Cristos and north across the Colorado line. Dustin lives in the southeast corner of the ranch, on a modest 60,000-acre section of shortgrass that, like the Tallgrass Preserve, supports bison. He spends a lot of his time restoring a semblance of what some call the prairie dog ecosystem.

The blue grama that dominates the prairie where Dustin works never stands very tall even in wet years. Fire recurs much less commonly than in the tallgrass prairie. Neither fire nor bison greatly alter the habitat value of the grass. But the black-tailed prairie dogs do. They are the base of the small-carnivore food chain and the grand architects of the shortgrass prairie. Some have called them "keystone" species because of the diversity of other animals they support.

Like the bison, prairie dogs convert grass to a lot of meat per acre. No other small animal does this nearly as well. Particularly important to the diversity equation, prairie dogs come in packages sized suitably for small meat-eaters. If you spend time on the Vermejo dogtowns, you'll see unusual numbers of bald eagles, golden eagles, ferruginous hawks, badgers, coyotes, swift foxes, prairie rattlesnakes, and others that eat prairie dogs.

No other small animal digs burrows as compulsively and as well as prairie dogs. An acre of dogtown may contain several hundred burrows. Deep and warm, flood-proof, more numerous than the prairie dogs in fact need, they make good havens for diverse species: burrowing owls, cottontails, snakes, lizards, toads, and others.

When Dustin came to Vermejo, prairie dogs lived in six colonies that in total covered about 500 acres. He transferred prairie dogs to new sites to start new colonies. Ten years later, the colonies numbered more than fifty and covered some 5,000 acres. Dogtown boarders and roomers proliferated proportionally.

Several years into the new millennium, Dustin's focus began to broaden to some of these other species. He measured the burgeoning population of burrowing owls. He released onto his prairie dog colonies that rarest of North American mammals and a prairie dog obligate, the black-footed ferret. Most important to the neighbors, he began developing plans to keep the prairie dog colonies from expanding beyond the boundaries of the Vermejo Park Ranch.

• • • • •

The hundredth meridian of longitude slices vertically through the Great Plains about halfway between the Tallgrass Prairie Preserve and the Vermejo Park Ranch. If you follow it south into Texas, you enter a country where sport hunting pays the costs of biodiversity management. Here, in the central and southern parts of the state, brush has proliferated in what used to be prairies and savannas. Juniper, scrub oak, and mesquite now choke landscapes that fire once kept open.

Texas has little public land, and many landless Texans like to hunt. Some lease private lands for the hunting rights. They show preference for central and southern Texas because white-tailed deer and exotics thrive especially well there, and the larger ranches provide that other ingredient that hunters crave: elbow room. Ranch owners cater to the hunters, who often pay a lot more per acre than the owner gets from running cows, sheep, or goats.

A strange crossbreed between redneck and artist now stalks this Texas land. No, not Willie Nelson and friends. These new professionals start their day at 6 a.m. in small-town coffee shops, then head out into the hinterlands to, as some put it, "sculpt brush."

The brush sculptors wield diesel-powered chisels from the driving-seats of great backhoes and clattering bulldozers. They take out rangeland brush in winding strips, irregular polygons, and more intricate patterns that conform with the lay of the land. They carve diversity from landscapes smothered by woody plants. Ranch owners and state grants programs pay for their work. A diverse throng of wildlife—deer, hoofed exotics of many species, turkeys, bobwhite quail, songbirds, small mammals—revels in their creations. The well-heeled hunters revel in the abundant wildlife and, perhaps, deep in their fatted Houston hearts, the longer view.

. . . . .

Maintenance of biodiversity on the Tallgrass Preserve, the Vermejo Park Ranch, the game ranches of Texas, and elsewhere out on the range costs money—money for burning grass, keeping prairie dogs off the neighbors, removing brush. Why should biodiversity—something that once occurred naturally—require money to maintain? Looking back to the time when diversity came free might give a clue.

Two to three centuries ago the Comanche, Apache, and other hunting cultures held sway over the southern grasslands. Beyond the wild food they gathered, the main energy sources they harnessed came in the form of horses and fire. And only fire had been among them more than a few generations. By hunting from horses and sometimes eating them, and by setting fires to improve horse and bison habitat, they increased their food supply.

But here their control ended. By necessity they bent to the forces they could not control: flood, drought, and population shifts and fluctuations in prey animals and food plants. Water, fire, and large grazers strafed the land with great irregularity in time and space. Great diversity resulted, often on grand spatial scales. The Indians moved about to find the best combinations.

Then came industrial-age Europeans. Their Bibles admonished control, and their taming of plants and animals and harnessing of vast new energies allowed it. They exercised to the maximum the power at their disposal. The ultimate result: small kingdoms overseen individually by small gods with a lot of control and the illusion of more. To demonstrate their control, the landlords imposed infrastructure and animals defiant of nature's diversifying agents. They brought water wells, croplands, fences, cows, goats, sheep, fire departments, dams. They instilled a cultural fear of losing control.

To free the agents of diversity now requires persistent effort. One must constrain fire, fence large grazers and poison small ones, and root out the woody plants resulting from constraint of grazers and fire. One must spend money.

Here and there, small voices question the wisdom of such compulsive control.

TWENTY Long Road Home

Figure 20. Rainbow's end. Photo by author.

My studies of wild animals and their habitats prompted me long ago to join the growing cadre that sees us humans near the limits of our own habitat. This belief arose not from religious conversion or teachings of a doomsday cult. It came from a few immutable principles familiar to habitat ecologists.

The logic runs like this. Habitats invariably have limits to the numbers of each kind of animal they can support. Symptoms of habitat limitation typically show up as starvation, malnutrition, disease, escalating social strife, or some combination thereof. No species, including ours, has proved capable of stabilizing or reducing its total numbers before running into the limits of habitat. Although some human populations indeed have limited their growth at lower levels, our total number keeps growing, and that matters in a world where people flow freely among regions.

We humans have proved ingenious at finding and applying subsidies to temporarily overcome the limits of our habitat. These aids make life better, at least for the moment. Who among us does not applaud life-saving advances in technology, new sources of energy, or food aid to hungry people? It is easier not to think about the consequences for our grandchildren: more people feeding on fewer resources.

The road home from this heedless journey is fraught with uncertainty. Like Odysseus, we need some reference points to guide us, some past experience to help us envision the future. For a start, let's assess the condition of the good ship Petroleum, the vessel that brought us so far along the last leg. How is this Promethean gift holding out?

Petroleum geologists come with the best expertise to answer these questions. Two who proved especially perceptive in this arena in the twentieth century worked for Shell Oil. Their names were M. King Hubbert and Kenneth Deffeyes.

Hubbert gained respect in the oil business when he foretold with uncanny accuracy the peak of oil production in America. An insider with Shell, in the 1950s he assembled all the data at his command, applied his own mathematical analyses, and predicted that oil production in the United States would peak in 1971. That year came and went, production numbers trickled in over the next several years and, lo, Hubbert had been off by a year or two at most.

Deffeyes spent his early years in the oil patches of Oklahoma and Texas. He later moved to the main Shell offices in Houston, where Hubbert reigned and in time became his mentor. Exceptionally bright, he eventually went on to teach at Princeton University. In 2001, from his position as professor emeritus at this stronghold of oil expertise, he published a book that applied Hubbert's methods to predict the world peak. According to his calculations, summed in the book titled *Hubbert's Peak*, we now sit very near the peak. It may take several more years of assessing production numbers to know whether Deffeyes was as prescient as his predecessor.

Why does peak oil production presage hard times for our species? Oil fuels a large part of the world's economy, and good substitutes for it don't exist, say Deffeyes and a host of others. So far, world oil production has risen with demand so as to keep prices moderate. Once production starts to decline, the gap between demand and supply will widen quickly; the prices of gasoline, diesel, and other derivatives of oil will skyrocket, and the costs of living the good life will follow suit.

Since the beginning of the twenty-first century, the oil industry seems to have exerted uncommonly strong control over the American federal government. The industry-government collaboration has drawn the country into an accelerated bid for other people's oil. The brighter among the scientists concerned with energy futures started long ago to consider other options. What substitutes for oil can propel tractors across the fields and cars and trucks down the highways at the rates to which we've become accustomed?

The fossil fuel options aren't too promising, said Ken Deffeyes in a follow-up to *Hubbert's Peak* he called *Beyond Oil*. Tar sands and oil shale may take us a little farther down the road, but they are going to be expensive. Then there's natural gas and coal gasification, neither anywhere near the cost-efficiency of oil. Regardless of the options eventually used, it will take a stupendous investment of energy and money to revamp the current production infrastructure. We're in a desperate situation, he concluded. It's like driving a car toward a cliff—once you're over the edge, it's a little late to hit the brakes.

Those positioned to benefit from selling biomass as energy are beginning to advertise their wares. Use ethanol! cries corporate agriculture. The corn belt can grow our fuel right here in the United States. Convert plant cellulose! trumpets another group. It makes more sense to look at the cornstalks than the corn grains. The Brazilians use bagasse, a cellulose-rich byproduct from sugarcane, to provide the energy for ethanol distillation from sugar. Use native grasses, recommended three scientists in the December 8, 2006, issue of *Science.* They planted diverse mixes—switchgrass and others—on abandoned agricultural lands and found the energy yield to increase with the number of species in the mix.

David and Marcia Pimentel of Cornell University, in their book *Food, Energy, and Society,* caution us not to count on a lot of energy from biomass. When you include all the energy costs of production, say they and their colleagues, neither corn grains nor sugarcane provide much net energy. In Brazil it takes seven acres of sugarcane to keep a small car running. This points to the auxiliary costs beyond the disappointing energy yield—soil erosion, water consumption, and loss of cropland for food production.

"How Green Are Biofuels?" So ran the title of an article in the January 2, 2008, issue of *Science.* Said authors Jörn Scharlemann and William Laurance: The careful analysts see a mountain of hidden environmental and economic costs no matter what the biofuel source. The cost ramifications are so complex, and comparisons of "apples and oranges" so common, that justifying one or any biofuel to provide even a small fraction of the energy now provided by oil comes with large uncertainties.

In the summer of 2008, oil billionaire T. Boone Pickens added his rhetoric to the agenda. "We're in trouble, folks," he declared. "We're running out of cheap oil."

Money speaks, and given the oil basis for T. Boone's wealth, presumably he knew wherewith he spoke. He turned up in Washington, giving testimony to Congress. "Invest in alternative energy sources such as wind and solar while we still have wealth enough to do so," he pleaded. Our representatives listened, or at least pretended to. Time will tell whether they can do anything effective and still retain favor with the voters.

. . . . .

Whatever substitutes for oil emerge, the inevitable rise of fuel costs will make a big difference out here in the lonely places. It's a long, long road to the feed stores and the livestock markets. People ranching for a living in the dry country already live on the economic edge. They have a lot of fence and water line to patrol, a collective fleet of fuel-hungry diesel pickups, and a lot of stock to truck back and forth. They will have a hard time reclaiming the cowboy heritage that rose to glory in the time before automobiles.

. . . . .

In 2006 I attended a meeting of people who preached that bison should replace cattle on western ranges. Most of the meeting attendees managed or worked on bison ranches. They saw the species as a sensible alternative to cattle for meat production. Bison thrived on coarser food than cattle, needed less supplementation in winter, and would travel farther from water to find forage. Not least to the meeting participants, the native bison seemed a more appropriate range icon than the European cow.

But, as with growing cattle, energy costs posed a major challenge. "Freight's eating us up," growled the manager of one large operation. "We've got to figure ways to cut shipping costs." "Good luck," Deffeyes might have said. "These are the good times."

Cutting costs would be hard. The bison industry had inherited an energy-intensive infrastructure, production mode, and marketing mentality bequeathed by the cattle industry. Most growers shipped their bison to be finished in cattle-type feedlots. Here the bison ate a cow-friendly grain ration that suited bison poorly to begin with, and that looked to get more expensive as the price of diesel climbed. From feedlots the bison rode eighteen-wheelers to often-distant processing plants, and from there the meat moved to even farther-flung consumer outlets.

Dave Carter, head of the National Bison Association, shared his vision of the future. Bison ranches would supply mostly local population centers as long-distance shipping to feedlots and distant markets became relics of the cheap-oil era. But how can we resolve the mismatch between human population distribution and that of bison? This will need some creative thinking.

Livestock producers are leaving the western rangelands in droves, said Kyran Kunkel of the World Wildlife Fund. They can't make a living the conventional way. Envision a future with wild bison herds roaming unfenced over a huge portion of this region. Something like a national park or grassland.

"My vision of the future," said Fred Dubray of the Cheyenne River Sioux tribe, "is a bunch of wild Indians chasing buffalo across the prairie on horseback." His hope smelled of process over product, a concept not well defined in the European pursuit of Manifest Destiny.

I had my own vision. Having hunted and eaten a lot of squirrels in my youth, I could see plains dwellers of future decades harvesting prairie dogs along with bison or cattle. Close cousin of the tree squirrel, prairie dogs taste much the same. Many native North Americans once ate them; some still do. My calculations show that a prairie dog colony could yield as much or more meat per acre than bison or cattle on the same range. They wouldn't require supplementation in winter, water in summer, or fences. Not least, their harvest and processing could bring kids back into the tradition of family sustenance. But I didn't broach my idea at the meeting.

I heard little mention of cattle at the meeting, either. Had cowboy prophets been there, they might have proposed a resurrection of the longhorn cow. Wild cow roundups and long drives to market might attract not only public acceptance but even participation. The agriculture-oriented television station RFD-TV sometimes features mini-versions of *Lonesome Dove.* Load up your horses in the gooseneck, drive on out, and join us in the trail ride to the future!

Cost efficiency counts a lot in capitalist enterprises. Pastoralists probably have been sensitive to costs ever since they herded the first goat. But they also have been influenced not a little by tradition, which in a changing market may not always make economic sense. The duel between profit and tradition will be interesting to follow as the cost of hauling stock up and down the highway in the Great American West climbs to unprecedented heights.

. . . . .

"The cowboy, an endangered species," proclaim T-shirts and wall placards in my county. Indeed, people born and raised somewhere else show up here in increasing numbers, and work-worn Wranglers and ratty boots seem less and less common. The wealthier among the newcomers buy old houses and renovate them, build new houses, or take up residence in rural subdivisions. Some wear clean Stetsons with fashionable crimps. Those without money look for places to park their hippie buses, or house-sit for those who do have money.

The misgivings that come with such changes have spawned organizations throughout the West dedicated to salvaging tradition. Some focus on what they call "sustainable ranching." The usual objectives seem to boil down to making an old lifeway viable in a new economy. Some organizations in my state—Malpais Borderlands Group, Quivira Coalition, Valles Calderas Trust—experiment with new ideas such as grass banking, ecotourism, federal ownership and management, and conservation easements. As nearly as I can judge, all get subsidized by outside money.

These groups face two basic challenges. First, agriculture in arid lands competes poorly with that in wetter lands. Such has always been the case. The Egyptian pharaohs got rich on farm production from the Nile Delta, not on goats and camels from the nearby deserts. The bulging billions in China and India subsisted over the centuries on rice from well-watered fields, not on sheep, horses, and yaks from steppe and mountain slope. Growing cattle in high-rainfall Florida and East Texas outcompetes that in New Mexico and Arizona hands down.

The second challenge likewise comes from the outside. Briefly it is that those willing to work for less outcompete those expecting more. In the modern world of easy international trade, this has brought ranching enterprises in poorer countries into direct competition with sustainable ranching in the American West. The Argentine gaucho brings down the world price of beef and the livelihood of the American cowboy simply by living on less.

The lesson for "sustainable" ranching in dry regions seems to be: get subsidized or sink. If the term *subsidy* carries a bad taste, from the indolence it sometimes generates or for other reasons, then call it something else. Many already do: jobs in town, drought relief, family inheritance, federal grants, ecotourism, social security.

I built my youthful dreams on the notion that I could find a remote homestead beyond a far mountain, grow some corn and beans, harvest deer and rabbits, and live off the land. And perhaps I could have, but not nearly in the comfort that would have appealed to the woman of my dreams. And not easily in competition with the throng of other men who came with similar visions. Arid lands ecology and too many people got in the way.

. . . . .

A cultural mismatch compounds the rising costs of ranching. Children of rural families follow the money to cities. Many who come from the cities to take their places lack the traditions—work habits, patience, tribal knowledge, subsistence ideals—to persevere without massive subsidy. Neither their home lives nor their formal educations prepared them for this.

David Schmidly, president of the University of New Mexico and previously head of Texas Tech and Oklahoma State universities in sequence, blames in part the universities. He explained his view in a 2005 article called "What It Means to Be a Naturalist, and the Future of Natural History at American Universities." With a scientist's tendency to understate, he calls the problem a "narrowing of focus" in education, as manifested in the withdrawal of students from field to laboratory, from real world to electronic image. I see the problem as abandonment of human habitat.

David and I briefly overlapped at Texas A&M in the early 1970s. He taught mammalogy in the tradition of the great old Texas naturalist William B. Davis. One legend about Davis claims that in summer, when he as professor took students to Mexico, he would drop each with a bag of traps and a bag of beans at a different location in the backcountry. His instructions: trap and prepare museum skins of local mammals, supplement your diet with mammal carcasses and other local foods as best you can, and I'll be back in two months to pick you up.

In 1961 Dr. Davis, then near retirement, helped me get my first professional job, a summer assignment with the Texas Game and Fish Department. By that time his rugged field courses, which he no longer headed, had retired into revered mythology. Still, he and others like him continued to shape the purpose of our university training. "You are here to learn," they would say. "You will spend a lot of time with us in the field, in the laboratory, and in the classroom. Your aim should be to protect a legacy, not to enrich yourselves. Those interested in making money are over there across campus in the Engineering Department."

Schmidly came from that heritage. Most universities don't emphasize field training anymore, he says. The pursuit of funding has proven infectious and diverting. Business ventures promising big bucks attract most of the better students. Pharmaceuticals, medicine, crop engineering, and related industries lead the way. To intercept the money stream, students jostle for place in laboratory and computer room.

This judgment matches my assessment of some of the younger university graduates I encounter. Most in the natural resources disciplines still advertise themselves as protectors of the environment, and some show great energy to this end. But a growing number seem drawn to a higher calling: Glorify Thyself. Maybe the advent of nature heroes and heroines on television has something to do with it. Getting on camera takes precedent over learning and serving.

According to Schmidly, the professors and universities bringing in the large grants look with arrogance and condescension on those of us who think students need outdoor instruction. They say it's not science. Many respond with disdain when we call for reconnecting urban kids with the outdoors.

Once advanced students get into the exhilarating game of big-stakes grants and public exposure, he says, their fascination with outdoor life atrophies. A new indoor tradition emerges. Their own kids grow up "clicking away on their plastic mice, happily viewing images of the very plants and animals they could be finding in the woods, streams, and meadows they no longer visit."

What does this portend for the future? I generally don't believe in conspiracy theories, but the rate at which the corporate-government coalition has taken control of our education—and thus our lives—disturbs me. The personal independence we inherited in America arose from economic independence. We still glorify the lifestyles—frontiersman, farmer, cowboy,

Indian—symbolizing that independence. But we have left them behind for a mess of pottage. With alarming speed we have come to depend almost totally on the corporate state to feed us. What would we do without Walmart?

We gave up the home-based training ground for independence when we moved to town. Now we're forsaking the formal education that could have provided a partial substitute. Big business and complicit governments applaud both. Their raw materials—the workers and consumers of America—respond obediently to the directives to work and to buy, directives that enrich the minority and whittle away at the security of the majority. The workers feel trapped; they see few options. They have learned no other way.

• • • • •

Leah Gray Jones runs an outfitting business from her home on Whitewater Mesa east of Alma, New Mexico. Vast public lands and the Gila Wilderness loom nearby. She knows horses, pack gear, and Dutch-oven cooking, and regularly takes visitors on day rides, weeklong pack trips, and hunting excursions. Born in Tucson, she moved at a young age with her family to Deming, New Mexico. She has spent most of her life in view of the Mogollon Mountains and the Gila, and more than a quarter century introducing people to the backcountry.

Says Leah: "My great-great-grandfather Reuben W. Gray came to New Mexico after the Civil War. He arrived driving a herd of cattle from the San Saba River country in Texas. Many of my ancestors are buried back there in the Gray Cemetery, near Cherokee."

In his book *The Longhorns,* J. Frank Dobie says Reuben Gray gathered 1,700 cattle on the San Saba in 1868 and trailed them west to Horsehead Crossing on the Pecos River. Not entirely coincidentally—there weren't a lot of trails west in those days—the better-known pioneers Charles Goodnight and Oliver Loving had passed over the same trail the previous year. Like Goodnight and Loving, Gray probably followed the Pecos north from the crossing, but unlike them, he stayed in New Mexico. Leah has deep family roots here.

In 2003, a college-prep school in Los Angeles bought a guest lodge several miles from Leah's home. For a couple of years, they carried on a novel experiment: bringing city-raised kids to learn rural lifeways and outdoor skills. Leah took the young people on backcountry trips and taught them how to ride and pack. She gave them adventures I suspect will last them a lifetime. But then, to at least my dismay, "The School" decided to abort the experiment, and the lodge went up for sale.

Leah doesn't make a lot of money, but she teaches lessons seldom available in more formal venues of education. So does Polly Tipton at her Double T Homestead, which offers bed, breakfast, and horseback rides a quarter mile down the road from me. And there are others. But the economic risks of these and similar "outdoor education" ventures weigh heavy against the meager profits.

In my view, these are the educational opportunities that deserve to be subsidized. Here stands the antidote to alienation and apathy in the generations to come. Here, in my fantasy, rises the New American University.

But then I climb the long Saturday ridge toward the mountain and see not a soul—not even tracks showing people have been there, except week-old ones of my own and those of Polly's horses. A hundred other slopes between here and Silver City likewise hear few boot-treads except from the middle-aged and older. This land belongs to the young but sees hardly any of them. Where are the students?

From my place on the ridge, I see a landscape with stories to tell. Nearby lies a millennium-old pithouse ruin with shards of pottery. To the north, west, and south jut landmarks commemorating fights between the last wild Apaches and first tame Americans a century and a quarter ago—White Rocks, Mineral Creek Canyon, Soldier Hill. Over there across the river stretches the WS Ranch, hangout in the late 1880s of train robber Butch Cassidy. Here and there are scattered purple glass chips and rusted lard cans, whispering of long-abandoned goat camps where less prosperous families of the mining days lived in tents and hauled firewood to the Mogollon mines on burros. Three generations of black plastic pipe, each generation larger in diameter than the last, snake down the slope from secret springs higher up. They end at the water tank near me.

The animals and plants lay out mysteries. What made those hand-sized tracks, the berry-filled scats, that nest of cactus and dried cow dung?

What's the story behind the ancient stumps of hand-hewn juniper, stumps outnumbered fifty-odd by the smaller living junipers now marching in and pushing back the grass? Charcoal marks the oldest stumps, bespeaking fires that once thinned trees but burn no more.

The grasses distribute themselves according to sun, shade, slope, soil, and past grazing by cow. The nutritious gramas—black, blue, sideoats, and hairy—grow each in its preferred place. The hardy curly-mesquite and vine-mesquite grasses thrive in the livestock-trampled open spaces. The grazing-sensitive bristlegrass and bush muhly hide beneath shrubs, away from the curling tongues of cows.

The brush-spattered slope falls away to Whitewater Creek down below. There, beside the road, the gaunt permittee loads his horse into the gooseneck trailer and heads home from a day of checking stock. Near at hand, the float on the empty water trough shows puncture marks from the teeth of bears. Tooth-marks likewise score the plastic pipes; bears have ripped one from its attachment, and the tank has no water. Someone has tied shut the valve with a plastic hay-string.

It's winter now, and the cattle find water at a spring nearby. The bears sleep in their dens. Come spring, they will emerge, the spring will dry up, the sun will heighten thirst among the rancher's cattle, and the endless task of keeping water coming down the mountain will resume.

A few miles to the north, Leah's small house sits on the north edge of a blue-grama mesa. Through binoculars I see her horses, like ants in the slanting sun, waiting for young riders. To the east, the mountain bluffs show gold. The sun goes down; quiet settles.

So many stories, no one to listen. Where have all the children gone? Down below and far away, to the bright seductive lights that pass for home.

Epilogue

A quarter century ago, ecologist Garret Hardin in his book *Filters Against Folly* lamented the inertia of cultural tradition. Resistance to social change, he observed, hinders people from adjusting to new circumstance. To face an uncertain future, he said, we need "far-reaching modifications of long-standing social arrangements." Such adjustments seldom come from within existing institutions or with their blessings, however. New ideas require outside thinking.

Traditions of grassland management, solidified as they are by isolation and by the power of the landscape to stiffen the psyche, resist change particularly well. The good, the bad, and the ugly of cowboy culture cling stubbornly to those indoctrinated in its rites. To explore which traditions are worth keeping and which are not, and to change accordingly, will require money from the outside and freedom from cultural censure.

One of the places I work in New Mexico offers these things. Regular cowboys work on the Armendaris Ranch, but influx of money from the outside eases the traditional obsession with meat production. Media billionaire Ted Turner, who acquired the ranch in 1994, provides funding. Manager Tom Waddell aims eventually to make the ranch self-sustaining—if he can—from running bison, offering big-game hunts, entertaining tourism, and pursuing other ventures. For now, the outside money makes room for creative thinking and experimentation.

Cattle no longer live on the Armendaris. As on all Turner ranches, bison have replaced them, and the cowboys have adapted. Ranch hands work with biologists of the Turner Endangered Species Fund to restore regionally scarce or extirpated species such as black-tailed prairie dogs, desert bighorn sheep, aplomado falcons, and bolson tortoises. Some of these efforts would attract a lot of ridicule on traditional ranches.

A large portion of the Armendaris remains devoid of bison or any animal managed as livestock. This departure from convention may spread inexorably to other arid ranches as subsidies for meat production dry up. Waddell, having spent his life participating in and observing southwestern ranching culture, knows that attempting to make money in conventional ways on such terrain carries extraordinary risk.

The Armendaris does a lot to educate young and old alike. It contributes funding to the local public schools and, more important in my view, brings people to the ranch. Groups from public schools and universities come for tours. University students conduct studies for advanced degrees. Evenings in late summer often see people from various walks of life watching several million bats pour from caves in the remote Armendaris Lava Field at the center of the ranch. The ranch helps renew a people-land connection that is crucial to our future but gets little notice in the conventional American system of education.

Some other ranches in New Mexico and the wider West are likewise beginning to challenge convention. Folks with incomes earned elsewhere own a disproportionate number of these. Most still run cattle—cowboy legends die hard. But in addition, some educate, some work to restore biodiversity, and some advertise wildlife or wilderness tours. Experimenting with nontraditional ways of living out on the range may be an idea whose time has come.

. . . . .

Born into a lower-income stratum of American society, I find comfort among people with modest means. Especially in rural places, those with little money show connections to habitat that resonate with some of my own experiences. But as people living from hand to mouth probably always have done, most I know who came from a subsistence lineage have little patience with conserving animals you can't eat. Indeed, absent government or other restrictions on exploitation, they often bring about some rendition of Garret Hardin's "tragedy of the commons," diminishing even the useful.

Wealthier people generally do better at conservation. They can afford the lost incomes, and live animals provide them more status than dead ones.

My reading of history leads to the same uneasy conclusion: the unregulated appetites and populations of ordinary people have exterminated, time and again, that which they needed and loved. In early Europe, only under the protection of the czars, princes, and liege lords did the large wild animals—aurochs, European bison, red deer, and others—survive the hungry mouths of the rabble. Robin Hood, a hero of my youth, probably would have hunted the king's deer to extinction given the chance. The same scenario played out in frontier America. Ordinary folks like Wright Mooar killed, to the limits of their technologies, the objects of their passions.

Unexpected riches in twentieth-century America stilled the hand of the human predator and brought wildlife back to lands private and public. As the twenty-first century advances, will a hungrier democracy on the landscape be able to preserve a proper human habitat with its attendant wild animals? Will we ordinary folk be able somehow to defy our age-old pattern of exploiting to the limits of our technologies? If we do not, our descendants—if indeed they have any appreciation for history—will look back upon these as the darkest of days.

When other animals overshoot the capacities of their habitats, the consequences for the generations that follow are not pretty. I find no logic that suggests we are different, that technology or favored status with the gods will let us somehow avoid retribution. By just about any reliable measure, colossal challenges lie ahead of us. They will command all the creative energies we can muster, out on the grass.

Notes

1. PROMETHEAN LEGACY

First atomic bomb: Walker 1995–2005
Humans as economic animals: Tiger and Fox 1971, Potts 1996, Whybrow 2005
Energy as source of wealth: Odum and Odum 1976, Johnson 1978
Juvenile dispersal: Horn 1978, Robinson and Bolen 1989, Gaines and Bertness 1993
True believer syndrome: Hoffer 1951

2. OUT OF THE FOREST

Grasses in human sustenance: deWet 1981

3. SCIENCE AND FAITH

Traditional range management paradigms: Weaver and Clements 1938, Dyksterhuis 1949, Stoddart and Smith 1955
Succession theory elaborated: Clements 1936, 1949; Weaver and Clements 1938; Dyksterhuis 1949

Succession theory criticized: Gleason 1939, Larson 1940, Egler 1954, Laycock 1989, Smith 1989, Borman and Pike 1994, Task Group on Unity in Concepts and Terminology 1995

Consequences of human overpopulation: Hardin 1980, 1985; Tainter 1988; Redman 1999; Diamond 2005

5. PLEASING TO THE EYE

Designing with nature: McHarg 1969

Human landscape preferences: Tuan 1974; Appleton 1975; Orians 1980; Balling and Falk 1982; Herzog 1984, 1987; Kaplan 1987; Herzog and Smith 1988; Heerwagen and Orians 1993; Hart and Sussman 2005

6. WHERE THE SHORT GRASS GROWS

Ancestral human habitats: Fagan 1990, Leakey 1994, Potts 1996, Hart and Sussman 2005

Evolution of hunting: Laughlin 1968, Washburn and Lancaster 1968, Leakey 1994, Potts 1996, Western 1997

Grazing lawns: McNaughton 1984, 1989

Sanctuary from stalking predators: Schaller 1972, Curio 1976, Sinclair et al. 2007

7. TURF

History of turf sports: Beard 1973, 1992; Jenkins 1994; Borman et al. 2001; Houston Croquet Association 2004; Wikipedia Online

History of lawns: Beard 1973, Jenkins 1994, Borman et al. 2001

Origin of turf and lawn grasses: Beard 1973, 1992; Jenkins 1994

Lawns as status: Jenkins 1994, Borman et al. 2001

Turfgrass species and varieties: Beard 1973

8. GRASSES AND GRAZERS: AN ECOLOGICAL PRIMER

Lewis and Clark journals: Biddle 1962, Moulton 1987

John Weaver's work: Weaver and Clements 1938; Weaver 1954, 1965

Grass response to drought: Albertson and Weaver 1946, Weaver 1954, Martin and Cable 1974, Risser 1985, Bock and Bock 2000

Grass response to grazing: Larson 1940, Weaver 1954, Martin and Cable 1974, Milchunas et al. 1988, Heitschmidt and Stuth 1991, Bock and Bock 2000, Holechek et al. 2006, Milchunas 2006

Coevolution of grasses and grazers: Clayton 1981, Stebbins 1981, Axelrod 1985, Risser 1985
Larger grazers digest coarse forage: Janis 1976, Owen-Smith 1988, Lott 2002

9. BISON PLAINS AND PRAIRIE DOGS

Lewis and Clark journals: Biddle 1962, Moulton 1987
Other early accounts: McDermott 1956, Goodman and Lawson 1995, Shaw and Lee 1997
Grazing effects on grass height and composition: Larson 1940, Ellison 1960, Hart and Hart 1997, Bock and Bock 2000, Truett 2003
Grazing and prairie dogs: Snell and Hlavachick 1980, Knowles 1986, Truett et al. 2001, Truett 2003

10. TAMING OF THE WEST

Lonesome Dove epic: McMurtry 1985
Charles Goodnight: Haley 1949
Texas Panhandle Indians and settlement: Haley 1949, Kenner 1969
Bison slaughter: Gard 1960, Roe 1970, Dary 1974, Isenberg 2000
Wright Mooar, bison hunter: Gard 1960, Isenberg 2000
Coming of cattle to the Great Plains: Webb 1931, Clawson 1950

11. PRODUCTION SCIENCE COMES TO THE RANGE

Settlement and early livestock grazing, Great Plains: Webb 1931, Clawson 1950, Box 2001
Succession theory in early conservation: Clements 1916, Weaver and Clements 1938, Dyksterhuis 1949, Weaver 1954
Brush encroachment onto rangelands: Buffington and Herbel 1965, Hennessey et al. 1983, Young and Sparks 1985, Bahre 1991, Gibbens et al. 1992, Dick-Peddie 1993, Miller 1999
Brush control: Hamilton et al. 2004

12. THE LAST PARIAH

Black-tailed prairie dog listed as threatened: Gober 2000
Early observations of prairie dogs: Merriam 1902, Bailey 1905, Mearns 1907, Drumm 1926, Gregg 1926, Oakes 2000
Early calls for prairie dog control: Merriam 1902, Bailey 1905

Grazing effects on early prairie dog distribution: Mead 1899, Hubbard and Schmitt 1984, Truett et al. 2001, Truett 2003
Prairie dog eradication: Hubbard and Schmitt 1984, Roemer and Forrest 1996, Oakes 2000, Forrest and Luchsinger 2006
History and effects of plague: Cully and Williams 2001, Cully et al. 2006
Black-tailed prairie dog history in Arizona: Mearns 1907, Alexander 1932, Oakes 2000

13. THE TROUBLE WITH LIVESTOCK

Humans as predators: Laughlin 1968, Ortega y Gassett 1972, Shepard 1973, Tuan 1979, Potts 1996
History of livestock domestication: Zuener 1963, Clutton-Brock 1987, Caras 1996
Rangelands as dry lands: Clutton-Brock 1987, Bulliet 1990, Caras 1996
Progenitors of livestock: Zuener 1963, Clutton-Brock 1987, Caras 1996
Problems with industrialized production of livestock: Steinfeld et al. 2006, Niman 2007

14. SUBSIDIZING JOHN WAYNE

Historical grazing effects, Great Basin: Young and Sparks 1985
Declining profitability of ranching: Young and Sparks 1985, Bartlett et al. 1989, Donahue 1999, Holechek 2001, Power and Barrett 2001, Holechek and Hawkes 2003
Ranching dependency on fossil fuels: Heitschmidt et al. 1996, Holechek 2006

15. COLLATERAL DAMAGE

Fire history, southwestern forests: Leopold 1924, Cooper 1960, Miller 1994
Influence of grazing on wildfire: Leopold 1924, Savage and Swetnam 1990, Miller 1999
Influence of fire on vegetation: Leopold 1924, Cooper 1960, Clayton 1981, Anderson 1990, Miller 1999, Bock and Bock 2000
Aboriginal burning of grasslands: Pyne 1982, Bahre 1991, Potts 1996, Pinchot et al 1998
Use of fire to manage grasslands: Collins and Wallace 1990, Scifres 2004, Vermeire et al. 2004

16. COWBOY

John Ford, filmmaker: Eder 2007, www.answers.com/topic/john-ford-filmmaker
American cowboy image: Webb 1931, Culley 1940, Dary 1981, Jordan 1981

Foreigners buy early American ranches: French 1928, Webb 1931
Early economics of ranching: French 1928, Webb 1931, Donahue 1999
Current economics of ranching: Bartlett et al. 1989, Holechek and Hawkes 1993, Donahue 1999, Holechek 2001, Power and Barrett 2001

17. RESURRECTION

Succession theory: Clements 1936, 1949; Dyksterhuis 1949, Young et al. 2001
Assembly theory: Gleason 1939, Foster et al. 1990, Young et al. 2001
Restoration concepts: Sprugel et al. 1991, Howe 1999, Ehrenfeld 2000, Hobbs and Harris 2001, Simpson 2002, Society for Ecological Restoration International Science & Policy Working Group 2004
Prairie dog biogeographic history: Goodwin 1995
Large grazers of the Pleistocene: Graham and Lundelius 1984, Martin 2005

18. PLEISTOCENE PARK

Restoration by a Pleistocene template: Donlan et al. 2005, Martin 2005
Charismatic megafauna: Donlan et al 2005, Hart and Sussman 2005, Martin 2005
First cattle to Texas: Dobie 1941, Truett and Lay 1984
Texas cattle empire and drives to market: Webb 1931, Dobie 1941, Haley 1949
Texas exotics: Mungall 2000, Schmidly 2002
Feral horses in early America: Wyman 1945
Ecological recovery of bison: Sanderson et al. 2008

19. DIVERSITY

Agricultural development and biodiversity: Smith 1976, deWet 1981, Swanson 1995
Human tendency toward specialization: deWet 1981, Campbell 1995
Managing rangelands for heterogeneity: Fuhlendorf and Engle 2001, 2004; Truett et al. 2001
Using fire to promote heterogeneity: Hamilton 1996, Fuhlendorf and Engle 2004
Prairie dogs promote diversity: Truett et al. 2001, 2006
Brush sculpting: Wiedemann 2004
Energy begets money: Odum and Odum 1976, Johnson 1978

20. LONG WAY HOME

Peak oil production: Deffeyes 2001, 2005
Energy from biofuels: Pimentel et al. 1996, Pimentel and Pimentel 1996, Tilman et al. 2006, Wald 2007, Scharlemann and Laurance 2008

Economics of conventional ranching: Holechek and Hawkes 1993, Power and Barrett 2001
Decline in natural history instruction: Schmidly 2005
Early cattle drives of Reuben Gray, Charles Goodnight, Oliver Loving: Dobie 1941, Haley 1949

EPILOGUE

Inertia of tradition: Hardin 1985
Tragedy of the commons: Hardin 1980
Historical protection of wildlife by the wealthy: Leopold 1933

References

Albertson, F. W., and J. E. Weaver. 1946. Reduction of ungrazed mixed prairie to short grass as a result of drought and dust. *Ecological Monographs* 16 (4): 449–63.
Alexander, A. M. 1932. Control, not extermination, of *Cynomys ludovicianus arizonensis*. *Journal of Mammalogy* 13: 302.
Anderson, R. C. 1990. The historic role of fire in the North American grassland. In S. L. Collins and L. L. Wallace, eds., *Fire in North American tallgrass prairies*. Norman: University of Oklahoma Press.
Appleton, J. 1975. *The experience of landscape*. New York: John Wiley and Sons.
Axelrod, D. I. 1985. Rise of the grassland biome, central North America. *Botanical Review* 51 (2): 163–210.
Bahre, C. J. 1991. *A legacy of change: Historic human impact on vegetation of the Arizona borderlands*. Tucson: University of Arizona Press.
Bailey, V. 1905. *Biological survey of Texas*. U.S. Department of Agriculture Bureau

of Biological Survey, North American Fauna No. 25. Washington, DC: U.S. Government Printing Office.

Balling, J. D., and J. H. Falk. 1982. Development of visual preference for natural environments. *Environment and Behavior* 14 (1): 5–28.

Bartlett, E. T., R. G. Taylor, J. R. McKean, and J. G. Hof. 1989. Motivation of Colorado ranchers with federal grazing allotments. *Journal of Range Management* 42: 454–57.

Beard, J. B. 1973. *Turfgrass: Science and culture.* Englewood Cliffs, NJ: Prentice-Hall.

———. 1992. The evolution of turfgrass sod. In J. L. Betts and W. G. Mathews, eds., *Turfgrass: Nature's constant benediction—the history of the American Sod Producers Association, 1967–1992.* Rolling Meadows, IL: American Sod Producers Association.

Biddle, N., ed. 1962. *The Journals of the Expedition under the Command of Capts. Lewis and Clark to the Sources of the Missouri, Thence across the Rocky Mountains and down the River Columbia to the Pacific Ocean, Performed during the years 1804–5–6 by Order of the Government of the United States.* Vol. 1. New York: Heritage Press.

Bock, C. E., and J. H. Bock. 2000. *The view from Bald Hill.* Berkeley: University of California Press.

Borman, M. M., and D. A. Pyke. 1994. Successional theory and the desired plant community approach. *Rangelands* 16 (2): 82–84.

Bormann, F. H., D. Balmori, and G. T. Geballe. 2001. *Redesigning the American lawn: A search for environmental harmony.* 2d ed. New Haven, CT: Yale University Press.

Box, T. 2001. From the dust of shame: A history on how the profession of range management was born. *Rangelands* 23 (6): 23.

Buffington, L. C., and C. H. Herbel. 1965. Vegetational changes on a semidesert grassland from 1858 to 1963. *Ecological Monographs* 35 (2): 139–64.

Bulliet, R. W. 1990. *The camel and the wheel.* New York: Columbia University Press.

Campbell, B. 1995. *Human ecology.* 2d ed. New York: Aldine de Gruyter.

Caras, R. A. 1996. *A perfect harmony: The intertwining lives of animals and humans throughout history.* New York: Simon and Schuster.

Clawson, M. 1950. *The western range livestock industry.* New York: McGraw-Hill.

Clayton, W. D. 1981. Evolution and distribution of grasses. *Annals of the Missouri Botanical Gardens* 68: 5–14.

Clements, F. E. 1916. *Plant succession.* Carnegie Institute Publication No. 42. Washington, DC: Carnegie Institute.

———. 1936. Nature and structure of the climax. *Journal of Ecology* 24: 252–84.

———. 1949. *Dynamics of vegetation.* New York: Wilson.

Clutton-Brock, J. 1987. *A natural history of domesticated animals.* Austin: University of Texas Press.

Collins, S. L., and L. L. Wallace. 1990. *Fire in North American tallgrass prairies.* Norman: University of Oklahoma Press.

Cooper, C. F. 1960. Changes in vegetation, structure, and growth of southwestern pine forests since white settlement. *Ecological Monographs* 30 (2): 129–64.

Culley, J. H. 1940. *Cattle, horses, and men.* Tucson: University of Arizona Press.

Cully, J. F., and E. S. Williams. 2001. Interspecific comparisons of sylvatic plague in prairie dogs. *Journal of Mammalogy* 82: 894–904.

Cully, J. F., D. E. Biggins, and D. B. Seery. 2006. Conservation of prairie dogs in areas with plague. In J. L. Hoogland, ed., *Conservation of the black-tailed prairie dog: Saving North America's western grasslands.* Washington, DC: Island Press.

Curio, E. 1976. *The ethology of predation.* Berlin: Springer-Verlag.

Dary, D. A. 1974. *The buffalo book: The saga of an American symbol.* New York: Avon.

———. 1981. *Cowboy culture: A saga of five centuries.* New York: Avon.

Deffeyes, K. S. 2001. *Hubbert's Peak: The impending world oil shortage.* Princeton, NJ: Princeton University Press.

———. 2005. *Beyond oil: The view from Hubbert's Peak.* New York: Hill and Wang.

deWet, J. M. J. 1981. Grasses and the culture history of man. *Annals of the Missouri Botanical Gardens* 68: 87–104.

Diamond, J. 2005. *Collapse: How societies choose to fail or succeed.* New York: Viking Press.

Dick-Peddie, W. A. 1993. *New Mexico vegetation: Past, present, and future.* Albuquerque: University of New Mexico Press.

Dobie, J. F. 1941. *The longhorns.* New York: Grossett and Dunlap.

Donahue, D. L. 1999. *The western range revisited: Removing livestock from public lands to conserve native biodiversity.* Norman: University of Oklahoma Press.

Donlan, J., et al. 2005. Re-wilding North America. *Nature* 436: 913–14.

Drumm, S. M., ed. 1926. *Down the Santa Fe Trail and into Mexico: The diary of Susan Shelby Magoffin, 1846–1847.* New Haven, CT: Yale University Press.

Dyksterhuis, E. J. 1949. Condition and management of range land based on quantitative ecology. *Journal of Range Management* 2: 104–15.

Eder, B. 2007. Biography: John Ford. *New York Times,* Jan. 19, 2007.

Egler, F. E. 1954. Vegetation science concepts 1: Initial floristic composition, a factor in old-field vegetation development. *Vegetatio* 4: 412–17.

Ehrenfeld, J. G. 2000. Defining the limits of restoration: The need for realistic goals. *Restoration Ecology* 8: 2–9.

Ellison, L. 1960. Influence of grazing on plant succession of rangelands. *Botanical Review* 26 (1): 1–78.

Fagan, B. M. 1990. *The journey from Eden: The peopling of our world.* London: Thames and Hudson.

Forrest, S. C., and J. C. Luchsinger. 2006. Past and current chemical control of prairie dogs. In J. L. Hoogland, ed., *Conservation of the black-tailed prairie dog: Saving North America's western grasslands.* Washington, DC: Island Press.

Foster, D. R., P. K. Schoonmaker, and S. T. A. Pickett. 1990. Insights from paleoecology to community ecology. *Trends in Ecology and Evolution* 5 (4): 119–22.

French, W. 1928. *Recollections of a western ranchman.* New York: Frederick A. Stokes. (Rpt. Silver City, NM: High Lonesome Books, 1990.)

Fuhlendorf, S. D., and D. M. Engle. 2001. Restoring heterogeneity on rangelands: Ecosystem management based on evolutionary grazing patterns. *BioScience* 51: 625–32.

———. 2004. Application of the fire-grazing interaction to restore a shifting mosaic on tallgrass prairie. *Journal of Applied Ecology* 41: 604–14.

Gaines, S. D., and M. Bertness. 1993. The dynamics of juvenile dispersal: Why field ecologists must integrate. *Ecology* 74: 2430–35.

Gard, W. 1960. *The great buffalo hunt: Its history and drama, and its role in the opening of the West.* New York: Alfred A. Knopf.

Gibbens, R. P., R. F. Beck, R. P. McNeely, and C. H. Herbel. 1992. Recent rates of mesquite establishment in the northern Chihuahuan Desert. *Journal of Range Management* 45: 585–88.

Gleason, H. A. 1939. The individualistic concept of the plant association. *American Midland Naturalist* 21: 92–110.

Gober, P. 2000. Endangered and threatened wildlife and plants: Twelve-month finding for a petition to list the black-tailed prairie dog as threatened. *Federal Register* 65 (24): 5476–88.

Goodman, G. J., and C. A. Lawson. 1995. *Retracing Major Stephen H. Long's 1820 expedition: The itinerary and botany.* Norman: University of Oklahoma Press.

Goodwin, H. T. 1995. Pliocene-Pleistocene biogeographic history of prairie dogs, genus *Cynomys (Sciuridae). Journal of Mammalogy* 76: 100–22.

Graham, R. W., and E. L. Lundelius, Jr. 1984. Coevolutionary disequilibrium and Pleistocene extinctions. In P. S. Martin and R. G. Klein, eds., *Quaternary extinctions: A prehistoric revolution.* Tucson: University of Arizona Press.

Gregg, J. 1926. *The commerce of the prairies.* Chicago: R. R. Donnelley and Sons. (Rpt. Lincoln: University of Nebraska Press, 1967.)

Haley, J. E. 1949. *Charles Goodnight: Cowman and plainsman.* Norman: University of Oklahoma Press.

Hamilton, R. G. 1996. Using fire and bison to restore a functional tallgrass prairie landscape. *Transactions of the 61st North American Wildlife and Natural Resources Conference.* Washington, DC: Wildlife Management Institute.

Hamilton, W. T., et al., eds. 2004. *Brush management: Past, present, future.* College Station: Texas A&M University Press.

Hardin, G. 1980. *Promethean ethics: Living with death, competition, and triage.* Seattle: University of Washington Press.

———. 1985. *Filters against folly: How to survive despite economists, ecologists and the merely eloquent.* New York: Penguin Books.

Hart, D., and R. W. Sussman. 2005. *Man the hunted: Primates, predators, and human evolution.* New York: Westview Press.

Hart, R. H., and J. A. Hart. 1997. Rangelands of the Great Plains before European settlement. *Rangelands* 19 (1): 4–11.

Heerwagen, J. H., and G. H. Orians. 1993. Humans, habitats, and aesthetics. In S. R. Kellert and E. O. Wilson, eds., *The biophilia hypothesis.* Washington, DC: Island Press.

Heitschmidt, R. K., R. E. Short, and E. E. Grings. 1996. Ecosystems, sustainability, and animal agriculture. *Journal of Animal Science* 74: 1395–1405.

Heitschmidt, R. K., and J. W. Stuth, eds. 1991. *Grazing management: An ecological perspective.* Portland, OR: Timber Press.

Hennessy, J. T., R. P. Gibbens, J. M. Tromble, and M. Cardenas. 1983. Vegetation changes from 1935 to 1980 in mesquite dunelands and former grasslands of southern New Mexico. *Journal of Range Management* 36: 370–74.

Herzog, T. R. 1984. A cognitive analysis of preference for field-and-forest environments. *Landscape Research* 9 (1): 10–16.

———. 1987. A cognitive analysis of preference for natural environments: Mountains, canyons, and deserts. *Landscape Journal* 6 (2): 140–52.

Herzog, T., and G. A. Smith. 1988. Danger, mystery, and environmental preference. *Environment and Behavior* 20 (3): 320–44.

Hobbs, R. J., and J. A. Harris. 2001. Restoration ecology: Repairing the earth's ecosystems in the new millennium. *Restoration Ecology* 9: 239–46.

Hoffer, E. 1951. *The true believer.* New York: Harper and Row.

Holechek, J. L. 2001. Western ranching at the crossroads. *Rangelands* 23 (1): 17–21.

———. 2006. Changing western landscapes, debt, and oil: A perspective. *Rangelands* 28 (1): 28–32.

Holechek, J. L., and J. Hawkes. 1993. Desert and prairie ranching profitability. *Rangelands* 15 (3): 104–9.

Holechek, J. L., T. T. Baker, J. C. Boren, and D. Galt. 2006. Grazing impacts on rangelands vegetation: What we have learned. *Rangelands* 28 (1): 7–13.

Horn, H. S. 1978. Optimal tactics of reproduction and life-history. In J. R. Krebs and N. B. Davies, eds., *Behavioural ecology: An evolutionary approach.* Sunderland, MA: Sinauer Associates.

Houston Croquet Association. 2004. *A brief history of croquet.* Houston: Houston Croquet Association.

Howe, H. F. 1999. Dominance, diversity, and grazing in tallgrass restoration. *Ecological Restoration* 17 (1–2): 59–66.

Hubbard, J., and G. Schmitt. 1984. *The black-footed ferret in New Mexico.* Santa Fe: New Mexico Department of Game and Fish.

Isenberg, A. C. 2000. *The destruction of the bison.* Cambridge: Cambridge University Press.

Janis, C. 1976. The evolutionary strategy of the Equidae and the origins of rumen and cecal digestion. *Evolution* 30: 757–74.

Jenkens, V. S. 1994. *The lawn: A history of an American obsession.* Washington, DC: Smithsonian Institution Press.

Johnson, W. 1978. *Muddling toward frugality.* San Francisco: Sierra Club Books.

Jordan, T. G. 1981. *Trails to Texas: Southern roots of western cattle ranching.* Lincoln: University of Nebraska Press.

Kaplan, S. 1987. Aesthetics, affect, and cognition: Environmental preference from an evolutionary perspective. *Environment and Behavior* 19 (1): 3–32.

Kenner, C. L. 1969. *A history of New Mexican–Plains Indians relations.* Norman: University of Oklahoma Press.

Knowles, C. J. 1986. Some relationships of black-tailed prairie dogs to livestock grazing. *Great Basin Naturalist* 46: 198–203.

Larson, F. 1940. The role of bison in maintaining the short grass plains. *Ecology* 21 (2): 113–21.

Laughlin, W. S. 1968. Hunting: An integrating biobehavior system and its evolutionary importance. In R. B. Lee and I. DeVore, eds., *Man the hunter.* New York: Aldine de Gruyter.

Laycock, W. A. 1989. Secondary succession and range condition criteria: Introduction to the problem. In W. K. Lauenroth and W. A. Laycock, eds., *Secondary succession and the evaluation of rangeland condition.* Boulder, CO: Westview Press.

Leakey, R. 1994. *The origin of humankind.* New York: Basic Books.

Leopold, A. 1924. Grass, brush, timber, and fire in southern Arizona. *Journal of Forestry* 22 (6): 1–10.

———. 1933. *Game management.* New York: Scribner.

Lott, D. F. 2002. *American bison: A natural history.* Berkeley: University of California Press.

Martin, P. S. 2005. *Twilight of the mammoths: Ice age extinctions and the rewilding of America.* Berkeley: University of California Press.

Martin, S. C., and D. R. Cable. 1974. *Managing semidesert grass-shrub ranges: Vegetation responses to precipitation, grazing, soil texture, and mesquite control.* U.S. Forest Service Technical Bulletin No. 1480. Washington, DC: U.S. Government Printing Office.

McDermott, J. F., ed. 1956. *Washington Irving: A tour on the prairies.* Norman: University of Oklahoma Press.

McHarg, I. L. 1969. *Design with nature.* New York: Wiley.

McMurtry, L. 1985. *Lonesome dove.* New York: Simon and Schuster.

McNaughton, S. J. 1984. Grazing lawns: Animals in herds, plant form, and coevolution. *American Naturalist* 124: 863–86.

McNaughton, S. J. 1989. Interactions of plants of the field layer with large herbivores. *Symposium of the Zoological Society of London* 61: 15–29.

Mead, J. R. 1899. Some natural history notes of 1859. *Transactions of the 30th and 31st annual meetings of the Kansas Academy of Science.* Topeka: J. S. Parks, State Printer.

Mearns, E. A. 1907. Mammals of the Mexican boundary of the United States. Part 1: Families *Didelphiidae* to *Muridae. Smithsonian Institution, United States National Museum Bulletin 56.* Washington, DC: U.S. Government Printing Office.

Merriam, C. H. 1902. The prairie dog of the Great Plains. In *Yearbook of the United States Department of Agriculture, 2001.* Washington, DC: U.S. Government Printing Office.

Milchunas, D. G. 2006. *Responses of plant communities to grazing in the southwestern United States.* U.S. Forest Service General Technical Report RMRS-GTR-169. Fort Collins, CO: U.S. Forest Service.

Milchunas, D. G., O. E. Sala, and W. K. Lauenroth. 1988. A generalized model of the effects of grazing by large herbivores on grassland community structure. *American Naturalist* 132: 87–106.

Miller, M. E. 1994. Historic vegetation change in the Negrito creek watershed, New Mexico. M.S. thesis, New Mexico State University.

———. 1999. Use of historic aerial photography to study vegetation change in the Negrito Creek watershed, southwestern New Mexico. *Southwestern Naturalist* 44 (2): 121–37.

Moulton, G. E., ed. 1987. *The journals of the Lewis and Clark expedition.* Vol. 3. Lincoln: University of Nebraska Press.

Mungall, E. C. 2000. Exotics. In S. Demarais and P. R. Krausman, eds. *Ecology and management of large mammals in North America.* Upper Saddle River, NJ: Prentice Hall.

Niman, N. H. 2007. Pig out. *New York Times,* March 14, 2007.

Oakes, C. L. 2000. History and consequences of keystone mammal eradication in the desert grasslands: The Arizona black-tailed prairie dog *(Cynomys ludovicianus arizonensis).* Ph.D. diss., University of Texas.

Odum, H. T., and and E. C. Odum. 1976. *Energy basis for man and nature.* New York: McGraw-Hill.

Orians, G. H. 1980. Habitat selection: General theory and applications to human behavior. In J. S. Lockard, ed., *The evolution of human social behavior.* New York: Elsevier.

Ortega y Gasset, J. 1972. *Meditations on hunting.* Trans. Howard Wescott. New York: Charles Scribner's Sons.

Owen-Smith, R. N. 1988. *Megaherbivores: The influence of very large body size on ecology.* Cambridge: Cambridge University Press.

Pimentel, D., and M. Pimentel, eds. 1996. *Food, energy, and society.* 2d ed. Niwot: University Press of Colorado.

Pimentel, D., et al. 1996. Biomass: Food versus fuel. In D. Pimentel and M. Pimentel, eds., Food, energy, and society. 2d ed. Niwot: University Press of Colorado.

Pinchot, G., C. Miller, and G. T. Frampton, Jr. 1998. *Breaking new ground.* Washington, DC: Island Press.

Potts, R. B. 1996. *Humanity's descent: The consequences of ecological instability.* New York: Avon Books.

Power, T. M., and R. N. Barrett. 2001. *Post-cowboy economics: Pay and prosperity in the new American West.* Washington, DC: Island Press.

Pyne, S. J. 1982. *Fire in America: A cultural history of wildland and rural fire.* Princeton, NJ: Princeton University Press.

Redman, C. L. 1999. *Human impact on ancient environments.* Tucson: University of Arizona Press.

Risser, P. G. 1985. Grasslands. In B. F. Chabot and H. A. Mooney, eds., *Physiological ecology of North American plant communities.* New York: Chapman and Hall.

Robinson, W. L., and E. G. Bolen. 1989. *Wildlife ecology and management.* 2d ed. New York: Macmillan.

Roe, F. G. 1970. *The North American buffalo: A critical study of the species in its wild state.* 2d ed. Toronto: University of Toronto Press.

Roemer, D. M., and S. C. Forrest. 1996. Prairie dog poisoning in northern Great Plains: An analysis of programs and policies. *Environmental Management* 20: 349–59.

Sanderson, E. W., et al. 2008. The ecological future of the North American bison: Conceiving long-term, large-scale conservation of wildlife. *Conservation Biology* 22 (2): 252–66.

Savage, M., and T. W. Swetnam. 1990. Early nineteenth-century fire decline following sheep pasturing in a Navajo ponderosa pine forest. *Ecology* 71: 2374–78.

Schaller, G. B. 1972. *The Serengeti lion: A study of predator-prey relations.* Chicago: University of Chicago Press.

Scharlemann, J. P. W., and W. F. Laurance. 2008. How green are biofuels? *Science* 319 (Jan. 4).

Schmidly, D. J. 2002. *Texas natural history: A century of change.* Lubbock: Texas Tech University Press.

———. 2005. What it means to be a naturalist and the future of natural history at American universities. *Journal of Mammalogy* 86: 449–56.

Scifres, C. J. 2004. Fire ecology and the progression of prescribed burning for brush management. In W. T. Hamilton, A. McGinty, D. N. Ueckert, C. W. Hanselka, and M. R. Lee, eds., *Brush management: Past, present, future.* College Station: Texas A&M University Press.

Shaw, J. H., and M. Lee. 1997. Relative abundance of bison, elk, and pronghorn on the southern plains, 1806–1857. *Plains Anthropologist* 42 (159), Memoir 29: 163–72.

Shepard, P. 1973. *The tender carnivore and the sacred game.* New York: Charles Scribner's Sons.

Simpson, T. B. 2002. An open approach to ecosystem change: Adopting a new paradigm for ecological restoration and management. *Ecological Restoration* 20 (3): 190–94.

Sinclair, A. R. E., et al. 2007. Long-term ecosystem dynamics in the Serengeti: Lessons for conservation. *Conservation Biology* 21 (3): 580–90.

Smith. E. L. 1989. Range condition and secondary succession: A critique. In W. K. Lauenroth and W. A. Laycock, eds., *Secondary succession and the evaluation of range condition.* Boulder, CO: Westview Press.

Smith, P. E. L. 1976. *Food production and its consequences.* 2d ed. Menlo Park, CA: Cummings.

Snell, G. P., and B. D. Hlavachick. 1980. Control of prairie dogs—the easy way. *Rangelands* 2 (6): 239–40.

Society for Ecological Restoration International Science and Policy Working Group. 2004. *The SER International primer on ecological restoration.* Tucson: Society for Ecological Restoration International.

Sprugel, D. G. 1991. Disturbance, equilibrium, and environmental variability: What is 'natural' vegetation in a changing environment? *Biological Conservation* 58: 1–18.

Stebbins, G. L. 1981. Coevolution of grasses and herbivores. *Annals of the Missouri Botanical Gardens* 68: 75–86.

Steinfeld, H., et al. 2006. *Livestock's long shadow: Environmental issues and options.* Rome: Food and Agriculture Organization of the United Nations.

Stoddart, L. A., and A. D. Smith. 1955. *Range management.* New York: McGraw-Hill.

Swanson, T. M. 1995. Uniformity in development and the decline of biological diversity. In T. M. Swanson, ed., *The economics and ecology of biodiversity decline: The forces driving global change.* Cambridge: Cambridge University Press.

Tainter, J. A. 1988. *The collapse of complex societies.* Cambridge: Cambridge University Press.

Task Group on Unity in Concepts and Terminology. 1995. New concepts for assessment of rangeland condition. *Journal of Range Management* 48: 271–82.

Tiger, L., and R. Fox. 1971. *The imperial animal.* New York: Holt, Rinehart, and Winston.

Tilman, D., J. Hill, and C. Lehman. 2006. Carbon-negative biofuels from low-input high-diversity grassland biomes. *Science* 314: 1598–1600.

Truett, J. C. 2003. Migrations of grassland communities and grazing philosophies in the Great Plains: A review and implications for management. *Great Plains Research* 13: 3–26.

Truett, J. C., and D. W. Lay. 1984. *Land of bears and honey: A natural history of East Texas.* Austin: University of Texas Press.

Truett, J. C., K. Bly-Honness, D. H. Long, and M. K. Phillips. 2006. Habitat restoration and management. In J. E. Roelle, B. J. Miller, J. L. Godbey, and D. E. Biggins, eds., *Recovery of the black-footed ferret: Progress and continuing challenges.* U.S. Geological Survey Scientific Investigations Report 2005–5293. Reston, VA: USGS.

Truett, J. C., M. Phillips, K. Kunkel, and R. Miller. 2001. Managing bison to conserve biodiversity. *Great Plains Research* 11: 123–44.

Tuan, Y. 1974. *Topophilia: A study of environmental perception, attitudes, and values.* New York: Columbia University Press.

———. 1979. *Landscapes of fear.* New York: Pantheon Books.

Vermeire, L. T., R. B. Mitchell, S. D. Fuhlendorf, and R. L. Gillen. 2004. Patch burning effects on grazing distribution. *Journal of Range Management* 57: 248–52.

Wald, M. L. 2007. Is ethanol for the long haul? *Scientific American* 296 (1): 42–49.

Walker, G. 1995–2005. *Trinity site: The first atomic test.* Albuquerque: U.S. Department of Energy National Atomic Museum.

Washburn, S. L., and C. S. Lancaster. 1968. The evolution of hunting. In R. B. Lee and I. DeVore, eds., *Man the hunter.* New York: Aldine de Gruyter.

Weaver, J. E. 1954. *North American prairie.* Lincoln: Johnsen Publishing Company.

———. 1965. *Native vegetation of Nebraska.* Lincoln: University of Nebraska Press.

Weaver, J. E., and F. E. Clements. 1938. *Plant ecology.* 2d ed. New York: McGraw-Hill.

Webb, W. P. 1931. *The Great Plains.* Lincoln: University of Nebraska Press.

Western, D. 1997. *In the dust of Kilimanjaro.* Washington, DC: Island Press.

Whybrow, P. C. 2005. *American mania: When more is not enough.* New York: W. W. Norton.

Wyman, W. D. 1945. *The wild horse of the West.* Lincoln: University of Nebraska Press.

Young, J. A., and B. A. Sparks. 1985. *Cattle in the cold desert.* Reno: University of Nevada Press.
Young, T. P., J. M. Chase, and R. T. Huddleston. 2001. Community succession and assembly: Comparing, contrasting, and combining paradigms in the context of ecological restoration. *Ecological Restoration* 19 (1): 5–18.
Zeuner, F. E. 1963. *A history of domesticated animals.* New York: Harper and Row.

Index

Text: 10/14 Palatino
Display: Univers Condensed and Bauer Bodoni
Compositor: Integrated Composition Systems
Indexer: Marcia Carlson
Printer: Sheridan Books, Inc.